Lecture Notes in Mathematics

A collection of informal reports and seminars
Edited by A. Dold, Heidelberg and B. Eckmann, Zürich

172

Yum-Tong Siu
University of Notre Dame, IN/USA
Günther Trautmann
Universität Frankfurt/Main, BRD

Gap-Sheaves and Extension of Coherent Analytic Subsheaves

Springer-Verlag
Berlin · Heidelberg · New York 1971

ISBN 3-540-05294-1 Springer-Verlag Berlin · Heidelberg · New York
ISBN 0-387-05294-1 Springer-Verlag New York · Heidelberg · Berlin

Offsetdruck: Julius Beltz, Weinheim/Bergstr.

Preface

The goal of these lecture notes is to provide an introduction to the theory of gap-sheaves and coherent subsheaf extension, which has been developed in the past few years by W. Thimm, the authors, and others.

The reader is assumed to be somewhat familiar with general sheaf theory and the function theory of several complex variables, as can be found for example in Godement's "Topologie algébrique et théorie des faisceaux" and Gunning-Rossi's "Analytic functions of several complex variables". Most of the required background material in sheaf theory and function theory is collected in §0. Some knowledge of elementary module theory and the theory of topological vector spaces is also assumed. Complete reference is given for results quoted but not proved in these lecture notes.

In §1 through §5 the structure of coherent analytic sheaves and their cohomology groups are discussed. The discussion includes gap-sheaves, sheaves of local cohomology, closedness of coboundary modules, and duality. In §6 through §10 coherent subsheaf extension is treated. The treatment presents the principal techniques and results of subsheaf extension.

In these lecture notes there are no new results, except possibly for some straightforward improvements of known ones. However, a number of proofs given here are new. At the end of these lecture notes there are some historical notes giving the origins of various theorems and their proofs.

Since this set of lecture notes is intended only as an introduction to the subject, no attempt is made to cover every known result in this field. The theory of extension of abstract coherent analytic sheaves is not treated at all, because the techniques required for abstract sheaf extension are of a nature completely different from those used for subsheaf extension.

During the preparation of these lecture notes the first author was partially supported by a grant from the National Science Foundation and the second author was a Fulbright Scholar visiting at the University of Notre Dame. We wish to express our thanks to the National Science Foundation, the Fulbright-Hays Program, and the University of Notre Dame for their generosity. Finally we wish to thank Bonnie Parsons for her excellent typing.

Table of contents

§0 Preliminaries

We recall some definitions and theorems in sheaf theory and the theory of complex spaces which we will use later on. For general sheaf theory we refer to [6] and for the theory of analytic sheaves and complex spaces we refer to [9] and [17].

(0.1) A subset A in a topological space X is called locally closed if there exists an open subset U of X such that A is a closed subset of U. If $\mathfrak{F}$ is a sheaf of abelian groups on X, we define $\Gamma_A(X, \mathfrak{F})$ as the subgroup of all elements of $\Gamma(U, \mathfrak{F})$ whose supports are contained in A. $\Gamma_A(X, \mathfrak{F})$ is independent of U. If

$$0 \to \mathfrak{F} \to \mathcal{C}^0 \to \mathcal{C}^1 \to \mathcal{C}^2 \to \dots$$

is the canonical (or any other) flabby resolution of $\mathfrak{F}$, we define the groups $H^i_A(X, \mathfrak{F})$ as the cohomology groups of the complex

$$0 \to \Gamma_A(X, \mathcal{C}^0) \to \Gamma_A(X, \mathcal{C}^1) \to \Gamma_A(X, \mathcal{C}^2) \to \dots$$

and call them the groups of local cohomology with supports in A and coefficients in $\mathfrak{F}$. In particular, we have

$$H^0_A(X, \mathfrak{F}) = \Gamma_A(X, \mathfrak{F})$$

and directly from the definition we have the following excision theorem: If Y is an open subset of X and A is a closed subset of Y, then

$$H^i_A(X, \mathfrak{F}) = H^i_A(Y, \mathfrak{F}).$$

If A itself is open in X, then

$$H_A^i(X, \mathcal{F}) = H^i(A, \mathcal{F}).$$

(0.2) If

$$0 \to \mathcal{F}' \to \mathcal{F} \to \mathcal{F}'' \to 0$$

is an exact sequence of sheaves of abelian groups and if $\mathcal{F}'$ is flabby, then the sequence

$$0 \to \Gamma_A(X, \mathcal{F}') \to \Gamma_A(X, \mathcal{F}) \xrightarrow{p} \Gamma_A(X, \mathcal{F}'') \to 0$$

is also exact. For, if

$$s'' \in \Gamma_A(X, \mathcal{F}'') \subset \Gamma(U, \mathcal{F}''),$$

we can find an $s \in \Gamma(U, \mathcal{F})$ such that $p(s) = s''$ and

$$s|U-A \in \Gamma(U-A, \mathcal{F}').$$

Hence there exists an $s' \in \Gamma(U, \mathcal{F}')$ whose restriction to U-A agrees with s and which satisfies $s - s' \in \Gamma_A(X, \mathcal{F})$ and $p(s-s') = s''$.

If in general $\mathcal{F}'$ is not flabby, then by a flabby resolution of the sequence and by applying the above result, we obtain an exact sequence

$$0 \to H_A^0(X, \mathcal{F}') \to H_A^0(X, \mathcal{F}) \to H_A^0(X, \mathcal{F}'') \to H_A^1(X, \mathcal{F}') \to \dots .$$

Moreover, for every flabby sheaf $\mathfrak{F}$ we have $H^i_A(X, \mathfrak{F}) = 0$ for $i \geq 1$.

(0.3) If A is a locally closed subset of X and A' is a closed subset of A, then A": = A - A' is also a locally closed subset of X. If $\mathfrak{F}$ is a sheaf of abelian groups on X, then the sequence

$$0 \to \Gamma_{A'}(X, \mathfrak{F}) \to \Gamma_A(X, \mathfrak{F}) \xrightarrow{p} \Gamma_{A''}(X, \mathfrak{F})$$

is exact and, if in addition $\mathfrak{F}$ is flabby, p is surjective. The first assertion is trivial. The second follows from the fact that p is the restriction map

$$\Gamma_A(U, \mathfrak{F}) \to \Gamma_A(U-A', \mathfrak{F}).$$

If

$$0 \to \mathfrak{F} \to \mathcal{C}^0 \to \mathcal{C}^1 \to \mathcal{C}^2 \to \cdots$$

is a flabby resolution of $\mathfrak{F}$, then we have the following complex of exact sequences

$$0 \to \Gamma_{A'}(X, \mathcal{C}^i) \to \Gamma_A(X, \mathcal{C}^i) \to \Gamma_{A''}(X, \mathcal{C}^i) \to 0$$

and obtain from it the following exact sequence

$$0 \to H^0_{A'}(X, \mathfrak{F}) \to H^0_A(X, \mathfrak{F}) \to H^0_{A''}(X, \mathfrak{F}) \to H^1_{A'}(X, \mathfrak{F}) \to \cdots .$$

If A is a closed subset of X, by considering the triple (A, X, X-A) we derive the exact sequence

$$0 \to H^0_A(X, \mathcal{F}) \to H^0(X, \mathcal{F}) \to H^0(X-A, \mathcal{F}) \to H^1_A(X, \mathcal{F}) \to H^1(X, \mathcal{F}) \to \cdots .$$

The groups $H^i_A(X, \mathcal{F})$ are exactly the obstructions for the restriction maps

$$H^i(X, \mathcal{F}) \to H^i(X-A, \mathcal{F})$$

to be surjective.

(0.4) Let A be a locally closed subset of a topological space X and $\mathcal{F}$ be a sheaf of abelian groups on X. Denote by $\mathcal{H}^i_A \mathcal{F}$ the sheaf defined by the presheaf which assigns to an open subset U the group $H^i_A(U, \mathcal{F})$ and assigns to the inclusion map of open subsets $V \hookrightarrow U$ the restriction map

$$H^i_A(U, \mathcal{F}) \to H^i_A(V, \mathcal{F}).$$

From (0.3) we derive the exact sequence

$$0 \to \mathcal{H}^0_{A'} \mathcal{F} \to \mathcal{H}^0_A \mathcal{F} \to \mathcal{H}^0_{A''} \mathcal{F} \to \mathcal{H}^1_{A'} \mathcal{F} \to \mathcal{H}^1_A \mathcal{F} \to \mathcal{H}^1_{A''} \mathcal{F} \to \mathcal{H}^2_{A'} \mathcal{F} \to \cdots ,$$

where A' is a closed subset of A and A" = A - A'.

If A is closed in X, we denote by $\mathcal{R}^i_A \mathcal{F}$ the sheaf defined by the presheaf

$$U \mapsto H^i(U-A, \mathcal{F}).$$

Since $\varinjlim$ is an exact functor and

$$\varinjlim_{U \ni x} H^i(U, \mathcal{F}) = 0,$$

by (0.3) we obtain the exact sequence

$$0 \to \mathcal{H}_A^0 \mathcal{F} \to \mathcal{F} \to \mathcal{R}_A^0 \mathcal{F} \to \mathcal{H}_A^1 \mathcal{F} \to 0$$

and the isomorphisms

$$\mathcal{R}_A^i \mathcal{F} \approx \mathcal{H}_A^{i+1} \mathcal{F} \quad (i \geqq 1).$$

(0.5) Lemma. If $\mathcal{F}$ is a flabby sheaf on a topological space X and A is a closed subset of X, then the sheaf $\mathcal{H}_A^0 \mathcal{F}$ is flabby.

Proof. If U is an open subset of X and

$$s \in \Gamma(U, \mathcal{H}_A^0 \mathcal{F}) = \Gamma_A(U, \mathcal{F}) \subset \Gamma(U, \mathcal{F}),$$

then

$$\tilde{s} := \begin{cases} 0 & \text{on } X-A \\ s & \text{on } U \end{cases}$$

extends s to $\Gamma(U \cup (X-A), \mathcal{F})$ and, since $\mathcal{F}$ is flabby, $\tilde{s}$ can be extended to an element $\hat{s}$ of $\Gamma(X, \mathcal{F})$. However,

$$\hat{s}|X - A = \tilde{s}|X - A = 0$$

and so

$$\hat{s} \in \Gamma(X, \mathcal{H}_A^0 \mathcal{F})$$

and $\hat{s}$ extends s. Q.E.D.

(0.6) Lemma. Let A be a closed subset of a topological space X and $\mathcal{F}$ be a sheaf of abelian groups on X. Let q be a non-negative integer and assume that

$$H^i(X, \mathcal{H}_A^j \mathcal{F}) = 0 \text{ for } j \leqq q \text{ and } i \geqq 1.$$

Then the natural homomorphisms

$$H_A^i(X, \mathcal{F}) \to \Gamma(X, \mathcal{H}_A^i \mathcal{F})$$

are isomorphisms for $i \leqq q + 1$.

Proof. Let

$$0 \to \mathcal{F} \to \mathcal{C}^0 \to \mathcal{C}^1 \to \mathcal{C}^2 \to \cdots$$

be the canonical flabby resolution of $\mathcal{F}$ on X. Then the groups $H_A^i(X, \mathcal{F})$ are the cohomology groups of the complex

$$0 \to \Gamma_A(X, \mathcal{C}^0) \overset{d^0}{\to} \Gamma_A(X, \mathcal{C}^1) \overset{d^1}{\to} \Gamma_A(X, \mathcal{C}^2) \overset{d^2}{\to} \cdots$$

and the sheaves $\mathcal{H}_A^i \mathcal{F}$ are the cohomology sheaves of the complex

$$0 \to \mathcal{H}^0_A \mathcal{C}^0 \xrightarrow{\delta^0} \mathcal{H}^0_A \mathcal{C}^1 \xrightarrow{\delta^1} \mathcal{H}^0_A \mathcal{C}^2 \xrightarrow{\delta^2} \cdots$$

since for the flabby sheaves $\mathcal{C}^i$ we have

$$\mathcal{H}^i_A \mathcal{C}^i = 0 \quad \text{for } i \geqq 1.$$

We define

$$Z^i = \text{Ker } d^i \quad , \quad B^i = \text{Im } d^{i-1},$$

$$\mathcal{Z}^i = \text{Ker } \delta^i \quad , \quad \mathcal{B}^i = \text{Im } \delta^{i-1},$$

and obtain the exact sequences

$$(*) \qquad 0 \to \mathcal{Z}^i \to \mathcal{H}^0_A \mathcal{C}^i \to \mathcal{B}^{i+1} \to 0 \qquad (i \geqq 0)$$

$$(**) \qquad 0 \to \mathcal{B}^i \to \mathcal{Z}^i \to \mathcal{H}^i_A \mathcal{F} \to 0 \qquad (i \geqq 1),$$

where $\mathcal{Z}^0 = \mathcal{H}^0_A \mathcal{F}$. Since

$$\Gamma(X, \mathcal{H}^0_A \mathcal{C}^i) = \Gamma_A(X, \mathcal{C}^i)$$

(which is valid for any sheaf), we obtain

$$Z^i = \Gamma(X, \mathcal{Z}^i).$$

We are going to prove by induction on i that for $0 \leqq i \leqq q$ we have

$$(\#)_i \qquad \begin{cases} H^k(X, \mathcal{Z}^i) = H^k(X, \mathcal{B}^{i+1}) = 0 & (k \geqq 1) \\ B^{i+1} = \Gamma(X, \mathcal{B}^{i+1}). \end{cases}$$

If $i < q$ and $(\#)_i$ is valid, we get from (**) the exact sequence

$$H^k(X, \mathcal{B}^{i+1}) \to H^k(X, \mathcal{Z}^{i+1}) \to H^k(X, \mathcal{H}_A^{i+1}\mathcal{F})$$

and so

$$H^k(X, \mathcal{Z}^{i+1}) = 0$$

by the assumption of the lemma since $i + 1 \leqq q$. By (*) and this result we obtain the exact sequence

$$0 \to Z^{i+1} \to \Gamma_A(X, \mathcal{C}^{i+1}) \to \Gamma(X, \mathcal{B}^{i+2}) \to 0$$

and hence

$$B^{i+2} = \Gamma(X, \mathcal{B}^{i+2}),$$

because B^{i+1} is the cokernel of the map

$$Z^{i+1} \to \Gamma_A(X, \mathcal{C}^{i+1}).$$

Moreover, (*) implies the exactness of the sequence

$$H^k(X, \mathcal{H}_A^0\mathcal{C}^{i+1}) \to H^k(X, \mathcal{B}^{i+2}) \to H^{k+1}(X, \mathcal{Z}^{i+1})$$

and, since $\mathcal{H}_A^0 \mathcal{C}^{i+1}$ is flabby,

$$H^k(X, \mathcal{B}^{i+2}) = 0.$$

Hence $(\#)_i$ implies $(\#)_{i+1}$ for $i < q$. $(\#)_0$ is proved in the same way, because $\mathcal{Z}^0 = \mathcal{H}_A^0 \mathcal{F}$. Hence $(\#)_i$ is valid for $i \leqq q$. Now in the commutative diagram

$$\begin{array}{ccccccc} 0 \to & B^i & \to & Z^i & \to & H_A^i(X, \mathcal{F}) & \to 0 \\ & \downarrow & & \downarrow & & \downarrow & \\ 0 \to & \Gamma(X, \mathcal{B}^i) & \to & \Gamma(X, \mathcal{Z}^i) & \to & \Gamma(X, \mathcal{H}_A^i \mathcal{F}) & \to 0 \end{array}$$

for $i \leqq q + 1$ the first two vertical homomorphisms are isomorphisms and so is the third since

$$H^1(X, \mathcal{B}^i) = 0 \quad \text{for } i \leqq q + 1.$$

Q.E.D.

(0.7) <u>Corollary</u>. Under the same assumptions as in Lemma (0.6) the following three conditions are equivalent.

(i) $\mathcal{H}_A^i \mathcal{F} = 0$ for $i \leqq q$.

(ii) For any open subset U of X, $H_A^i(U, \mathcal{F}) = 0$ for $i \leqq q$.

(iii) For any open subset U of X the restriction maps

$$H^i(U, \mathcal{F}) \to H^i(U-A, \mathcal{F})$$

are bijective for $i < q$ and injective for $i = q$.

Proof. (ii) <=> (iii) follows from (0.3). (ii) => (i) is trivial. (i) => (ii) follows from Lemma (0.6). Q.E.D.

(0.8) Cartan's theorems A and B.

We assume as known the general theory of complex spaces and the notion of coherent sheaves (see [7], [9], and [17]). Unless specified otherwise, all complex spaces are considered unreduced (i.e. their structure sheaves may admit nonzero nilpotent elements). The notion of Stein spaces can be found in [7] and [9]. Cartan's theorems A and B are as follows.

If $(X, \mathcal{O})$ is a Stein space and $\mathcal{F}$ is a coherent analytic sheaf on X, then

(A) the image of the canonical homomorphisms

$$\Gamma(X, \mathcal{F}) \to \mathcal{F}_x$$

generates $\mathcal{F}_x$ over $\mathcal{O}_x$ for every $x \in X$, and

(B) $H^i(X, \mathcal{F}) = 0$ for $i \geqq 1$.

A proof of the reduced case can be found in [9] and the unreduced case is derived from the reduced case in [7].

(0.9) Leray's theorem. If X is a complex space and $\mathfrak{U} = \{U_\alpha\}_{\alpha \in A}$ is a covering of X such that each U_α is a Stein open subset of X, then for any coherent analytic sheaf $\mathcal{F}$ on X there are canonical isomorphisms

$$H^i(\mathfrak{U}, \mathcal{F}) \overset{\approx}{\to} H^i(X, \mathcal{F})$$

for $i \geq 0$ (where $H^i(\mathfrak{U}, \mathcal{F})$ are the Čech cohomology groups of $\mathcal{F}$ for the covering $\mathfrak{U}$).

(0.10) Hilbert Nullstellensatz.

Let $(X, \mathcal{O})$ be a complex space. If $f \in \Gamma(X, \mathcal{O})$, then to f there is assigned a continuous complex-valued function on X:

$$x \to \overline{f}_x ,$$

where $\overline{f}_x$ is the residue class of f_x in $\mathcal{O}_x / \mathfrak{m}_x$, $\mathfrak{m}_x$ being the maximal ideal in $\mathcal{O}_x$. We write $f(x) = \overline{f}_x$. If $\mathcal{A} \subset \mathcal{O}$ is a coherent ideal-sheaf, then

$$A = \operatorname{Supp} \mathcal{O}/\mathcal{A} = \{x \in X \mid f(x) = 0 \text{ for all } f_x \in \mathcal{A}_x\}$$

is the (analytic) subvariety defined by $\mathcal{A}$, which we denote by $V(\mathcal{A})$. If conversely A is a subvariety in X, then by

$$U \mapsto \mathcal{A}(U) = \{f \in \Gamma(U, \mathcal{O}) \mid f(x) = 0 \text{ for all } x \in A \cap U\}$$

there is defined an ideal-sheaf $\mathcal{A} = \mathcal{J}(A) \subset \mathcal{O}$. This sheaf $\mathcal{J}(A)$ is coherent by a theorem of H. Cartan (see e.g. [17, p. 77, Th. 5]; the coherence of $\mathcal{J}(A)$ in the case of a subvariety in $\mathbb{C}^n$ implies the coherence in the general case by taking quotients). It is easy to see that we always have

$$V(\mathcal{J}(A)) = A.$$

The Hilbert Nullstellensatz states that

$$\text{(HN)} \qquad \mathcal{J}(V(A)) = \operatorname{Rad} \mathcal{A}$$

where Rad $\mathcal{A}$ is the radical of $\mathcal{A}$ (i.e. Rad $\mathcal{A}$ is the ideal-sheaf defined by the presheaf

$$U \mapsto \operatorname{Rad} \mathcal{A}(U)$$

or equivalently the ideal-sheaf whose stalk at x is the radical of $\mathcal{A}_x$).

A proof can be found in [17, p. 43, Th. 2] (only the case of subvarieties in $\mathbb{C}^n$ is proved there, but the general case follows by writing $\mathcal{A}$ locally as a quotient of an ideal-sheaf in $\mathbb{C}^n$).

Suppose $\mathcal{A}$ and $\mathcal{B}$ are two coherent ideal-sheaves on X. If $V(\mathcal{A}) \subset V(\mathcal{B})$, then for any compact subset K of X there exists a natural number n such that $\mathcal{B}^n|K \subset \mathcal{A}|K$, because

$$\operatorname{Rad} \mathcal{B} = \mathcal{J}(V(\mathcal{B})) \subset \mathcal{J}(V(\mathcal{A})) = \operatorname{Rad} \mathcal{A}.$$

(0.12) If $(X, \mathcal{O})$ is a complex space and

$$\mathcal{F}_1 \subset \mathcal{F}_2 \subset \mathcal{F}_3 \subset \ldots \subset \mathcal{F}$$

is an ascending chain of coherent analytic subsheaves of a coherent analytic sheaf $\mathcal{F}$, then this chain is locally stationary, i.e. for every compact subset K of X there exists a natural number n depending on K such that

$$\mathcal{F}_m = \mathcal{F}_n \quad \text{for } m \geq n$$

(see [7, §1, Satz 8] or [23 , Prop. 1]).

(0.13) Topologies on cohomology groups.

Let X be a complex space with countable topology and $\mathcal{F}$ be a coherent analytic sheaf on X. If U is an open subset of X and is embedded as a subvariety in some open subset D of $\mathbb{C}^n$, then $\mathcal{F}|U$ can be considered as a coherent $\mathcal{O}_D$-sheaf with support in U, where $\mathcal{O}_D$ is the sheaf of germs of holomorphic functions on D. For every point x of U we can find a system of neighborhoods V of x of the form $V = U \cap G$ where G is a relatively compact Stein open subset of D. By Cartan's theorems A and B there is an exact sequence

$$\Gamma(G, \mathcal{O}_D^p) \xrightarrow{\alpha} \Gamma(G, \mathcal{O}_D^q) \xrightarrow{\beta} \Gamma(V, \mathcal{F}) \to 0$$

of $\Gamma(G, \mathcal{O}_D)$-homomorphisms. Since $\Gamma(G, \mathcal{O}_D^q)$ is a Fréchet space in the topology of uniform convergence on compact subsets, we can impose on $\Gamma(V, \mathcal{F})$ the quotient topology. It is well known that Im α is closed ([2] or [15]; cf. [9 , p. 85, II. D. 3]). Hence, $\Gamma(V, \mathcal{F})$ is a Fréchet space. This topology depends on neither the homomorphism α nor the imbedding of U into D.

If U is any open subset of X, then we can find a countable covering $\mathcal{V} = \{V_\alpha\}$ of U such that each V_α can be embedded as a subvariety of an open subset of some $\mathbb{C}^{n_\alpha}$. Each $\Gamma(V_\alpha, \mathcal{F})$ has the topology described above. The restriction maps

$$\Gamma(U, \mathcal{F}) \to \Gamma(V_\alpha, \mathcal{F})$$

induce an injective map

$$\Gamma(U, \mathfrak{F}) \to \Pi_\alpha \, \Gamma(V_\alpha, \mathfrak{F})$$

whose image is a closed subset of $\Pi_\alpha \, \Gamma(V_\alpha, \mathfrak{F})$. This gives a Fréchet space topology on $\Gamma(U, \mathfrak{F})$. This topology is independent of the covering $\mathfrak{V}$.

If $\mathfrak{U} = \{U_\alpha\}$ is a countable Stein covering of X, then

$$C^q(\mathfrak{U}, \mathfrak{F}) = \Pi \, \Gamma(U_{\alpha_0 \ldots \alpha_q}, \mathfrak{F})$$

can be equipped with the product topology and becomes a Fréchet space. The coboundary maps

$$\partial^q: \; C^q(\mathfrak{U}, \mathfrak{F}) \to C^{q+1}(\mathfrak{U}, \mathfrak{F})$$

are continuous and so

$$Z^q(\mathfrak{U}, \mathfrak{F}) = \mathrm{Ker} \; \partial^q$$

is a closed subspace. However, in general

$$B^q(\mathfrak{U}, \mathfrak{F}) = \mathrm{Im} \; \partial^{q-1}$$

is not closed. By Leray's theorem,

$$H^q(X, \mathfrak{F}) = Z^q(\mathfrak{U}, \mathfrak{F})/B^q(\mathfrak{U}, \mathfrak{F})$$

and hence $H^q(X, \mathfrak{F})$ can be given the quotient topology. This topology is independent of $\mathfrak{U}$. $H^q(X, \mathfrak{F})$ is a Fréchet space if and only if $B^q(\mathfrak{U}, \mathfrak{F})$ is closed in $Z^q(\mathfrak{U}, \mathfrak{F})$.

For two open subsets $V \subset U$ of X the restriction map

$$H^q(U, \mathfrak{F}) \to H^q(V, \mathfrak{F})$$

is continuous for $q \geqq 0$.

If

$$0 \to \mathfrak{F}' \to \mathfrak{F} \to \mathfrak{F}'' \to 0$$

is an exact sequence of coherent analytic sheaves on X, then in the cohomology sequence

$$0 \to H^0(X, \mathfrak{F}') \to H^0(X, \mathfrak{F}) \to H^0(X, \mathfrak{F}'') \to H^1(X, \mathfrak{F}') \to \dots$$

all homomorphisms are continuous. To prove it, one need only consider the Čech complexes $C^\cdot(\mathfrak{U}, \mathfrak{F})$ for a countable Stein covering $\mathfrak{U}$.

(0.14) Frenkel's lemma.

Let $0 \leqq \varepsilon_i < 1$ be real numbers $(1 \leqq i \leqq n)$ and let

$$U_i = \{(z_1, \dots, z_n) \in \mathbb{C}^n \mid \varepsilon_i < |z_i| < 1,\ |z_j| < 1 \text{ for } j \neq i\}$$

and

$$V^{(k)} = \{(z_1, \dots, z_n) \in \mathbb{C}^n \,\big|\, |z_i| < \varepsilon_i \text{ for } 1 \leqq i \leqq k \text{ and } |z_j| < 1 \text{ for } j > k\}$$

Suppose $\varepsilon_i > 0$ for $1 \leqq i \leqq k$. Introduce the open set

$$Q^{(k)} = V^{(k)} \cup U_{k+1} \cup \dots \cup U_n.$$

By Δ we denote the unit polydisc

$$\{(z_1,\dots,z_n) \in \mathbb{C}^n \,\big|\, |z_i| < 1 \text{ for } 1 \leqq i \leqq n\}$$

and we define

$$H^{(k)} = U_{k+1} \cup \dots \cup U_n.$$

Since

$$\mathfrak{U} := \{V^{(k)}, U_{k+1}, \dots, U_n\}$$

is a Stein covering of $Q^{(k)}$, for any coherent analytic sheaf $\mathcal{F}$ on $Q^{(k)}$ we have

$$H^i(Q^{(k)}, \mathcal{F}) = H^i(\mathfrak{U}, \mathcal{F})$$

and

$$H^i(Q^{(k)}, \mathcal{F}) = 0 \quad \text{for } i > n - k.$$

Frenkel's lemma states that, for the sheaf ${}_n\mathcal{O}$ of germs of holomorphic functions on $\mathbb{C}^n$,

(F_1) the restriction map $H^0(\Delta, {}_n\mathcal{O}) \to H^0(Q^{(k)}, {}_n\mathcal{O})$ is an isomorphism for $n \geqq 2$ and $H^i(Q^{(k)}, {}_n\mathcal{O}) = 0$ for $i \neq 0, n-k$;

(F_2) the restriction map $H^0(\Delta, {}_n\mathcal{O}) \to H^0(H^{(k)}, {}_n\mathcal{O})$ is an isomorphism for $k \leqq n - 2$ and $H^i(H^{(k)}, {}_n\mathcal{O}) = 0$ for $i \neq 0, n-k-1$.

Proofs of these facts can be found in [4], [1 , p. 218] or easily obtained by considering Laurent series expansions of the cocycles.

(0.15) Analytic cover [9 , p. 101]

Let U be a domain in $\mathbb{C}^n$ and A be a subset of U. A is said to be negligible if it is nowhere dense and if, for every subdomain U' of U and every function f holomorphic on U' - A and locally bounded on U', f has a unique holomorphic extension to all of U'.

An analytic cover is a triple (X,π,U) such that

(i) X is a locally compact Hausdorff space,

(ii) U is a domain in $\mathbb{C}^n$,

(iii) π is a proper continuous map from X to U such that $\pi^{-1}(y)$ is discrete for $y \in U$,

(iv) there are a negligible subset A of U and a natural number λ such that π is a λ-sheeted covering map from $X - \pi^{-1}(A)$ onto $U - A$,

(v) $X - \pi^{-1}(A)$ is dense in X.

We call A the critical set. By an abuse of language we also say that $\pi: X \to U$ is an analytic cover or X is an analytic cover over U under the projection π.

Suppose X is a subvariety of an open subset of $\mathbb{C}^N$. (When a subvariety is regarded as a complex space, it is always given the reduced complex structure unless specified otherwise.) Suppose U is a domain in $\mathbb{C}^n$ and $\pi: X \to U$ is a holomorphic map. Then (X,π,U) is an analytic cover if and only if X has pure dimension n and π is proper. The "only if" part is clear. For the "if" part, we observe that every compact subvariety of $\mathbb{C}^N$ is finite and take the critical set A to be $\pi(S \cup T)$, where S is the set of all singular points of X and T is the subset of X - S at which the holomorphic map $\pi|X-S$ has rank $< n$.

(0.16) Maximum principle

Suppose V is a connected subvariety of a domain U in $\mathbb{C}^n$. We have the following maximum principles.

(a) If f is a holomorphic function on U and $f|V$ attains its maximum modulus on V, then f is constant on V [9, p. 106, III. B. 16].

(b) If φ is a continuous plurisubharmonic function on U and $\varphi|V$ attains its maximum, then φ is constant on V [9 , p. 272, IX. C. 3].

It follows that, on a connected reduced complex space, a non-constant holomorphic function cannot attain its maximum modulus and a non-constant continuous plurisubharmonic function (i.e. a function which is locally the restriction of a continuous plurisubharmonic function on a domain in some complex number space when the complex space is locally embedded as a subvariety in the domain) cannot attain its maximum.

(0.17) Suppose V is a subvariety of an open neighborhood U of 0 in $\mathbb{C}^n$ and the dimension of V at 0 is $\geqq k$. If f is a holomorphic function on U and vanishes at 0, then the dimension of $V' := \{x \in V | f(x) = 0\}$ at 0 is $\geqq k - 1$. If $\dim_0 V = k$ and f does not vanish identically on any k-dimensional branch-germ of V at 0, then $\dim_0 V' = k-1$ [9 , p. 115, III. C. 14].

(0.18) Lemma. Suppose V is a subvariety of an open neighborhood of 0 in $\mathbb{C}^n$ and $\dim_0 V \leqq k$. Suppose $A = (a_{ij})$ is a nonsingular $n \times n$ matrix and $\varepsilon > 0$. Then there exists a nonsingular $n \times n$ matrix $B = (b_{ij})$ such that $|b_{ij} - a_{ij}| < \varepsilon$ and $\dim V \cap \{w_1 = \dots = w_k = 0\} = 0$, where $w_i = \Sigma_{j=1}^n b_{ij} z_j$ and $z_1, \dots, z_n$ are the coordinates of $\mathbb{C}^n$.

Proof. We can assume that ε is so small that, if an $n \times n$ matrix $B = (b_{ij})$ satisfies $|b_{ij} - a_{ij}| < \varepsilon$, then B is nonsingular.

Use induction on k. Let $\{V_\alpha\}$ be the set of all k-dimensional branches of V. Take $(z_1^{(\alpha)}, \dots, z_n^{(\alpha)}) \in V_\alpha$. We can choose $c_1, \dots, c_n \in$ such that $|c_j - a_{kj}| < \varepsilon$ and

$$\Sigma_{j=1}^n c_j z_j^{(\alpha)} \neq 0 \quad \text{for all } \alpha,$$

because

$$\bigcup_\alpha \{(\zeta_1, \dots, \zeta_n) \in \mathbb{C}^n \big| \Sigma_{j=1}^n \zeta_j z_j^{(\alpha)} = 0\}$$

is a countable union of subvarieties of dimension n-1 in $\mathbb{C}^n$. Apply the induction hypothesis to the subvariety

$$V \cap \{\Sigma_{j=1}^n c_j z_j = 0\}$$

which has dimension $\leqq k - 1$ by (0.17). We obtain a nonsingular $n \times n$ matrix $B' = (b'_{ij})$. Set

$$\begin{cases} b_{ij} = b'_{ij} & \text{for } 1 \leqq i \leqq k-1 \\ b_{kj} = c_j & \\ b_{ij} = a_{ij} & \text{for } k+1 \leqq i \leqq n. \end{cases}$$

Then $B = (b_{ij})$ satisfies the requirement. Q.E.D.

(0.19) Lemma. Suppose V is a subvariety of an open neighborhood U of 0 in $\mathbb{C}^n$ and $F: V \to \mathbb{C}^k$ is a holomorphic map such that $\dim_0 F^{-1}F(0) = 0$. Then there exists an open neighborhood U' of 0 in U such that $\dim_x F^{-1}F(x) = 0$ for $x \in U'$.

Proof. Since $\dim_0 F^{-1}F(0) = 0$, there exists a relatively compact open neighborhood W of 0 in U such that the boundary ∂W of W is disjoint from $\dim_0 F^{-1}F(0)$. Hence

$$F(\partial W) \cap F(0) = \emptyset.$$

Since $\mathbb{C}^k - F(\partial W)$ is an open neighborhood of $F(0)$, there exists an open neighborhood U' of 0 in W such that

$$F(U') \subset \mathbb{C}^k - F(\partial W).$$

Hence, for $x \in U'$,

$$F^{-1} F(x) \cap W = F^{-1} F(x) \cap W^{-}$$

is compact and $\dim_x F^{-1} F(x) = 0$. Q.E.D.

(0.20) Suppose $(X, \mathcal{O})$ and Y are complex spaces and $\pi: X \to Y$ is a holomorphic map (i.e. a morphism of ringed spaces). If $\mathcal{F}$ is an analytic sheaf on X, then the qth direct image $\pi_q(\mathcal{F})$ of $\mathcal{F}$ under π is the analytic sheaf on Y defined by the following presheaf:

$$U \to H^q(\pi^{-1}(U), \mathcal{F}).$$

If π is proper and has finite fibers and $\mathcal{F}$ is coherent, then the zeroth direct image $\pi_0(\mathcal{F})$ is a coherent analytic sheaf on Y. (For the case where X is reduced, this statement is proved in [17, p. 81, Th. 7]. For the general case, since π is proper, we can assume that $\mathcal{K}^k = 0$ for some integer k, where $\mathcal{K}$ is the subsheaf of all nilpotent elements of $\mathcal{O}$. Since π has finite fibers, by Cartan's theorem B

$$0 \to \pi_0(\mathcal{K}\mathcal{F}/\mathcal{K}^{\ell}\mathcal{F}) \to \pi_0(\mathcal{F}/\mathcal{K}^{\ell}\mathcal{F}) \to \pi_0(\mathcal{F}/\mathcal{K}\mathcal{F}) \to 0$$

is exact. $\mathcal{K}\mathcal{F}/\mathcal{K}^{\ell}\mathcal{F}$ and $\mathcal{F}/\mathcal{K}\mathcal{F}$ can be regarded respectively as coherent analytic sheaves on $(X, \mathcal{O}/\mathcal{K}^{\ell-1})$ and $(X, \mathcal{O}/\mathcal{K})$. Hence the general case follows from induction on k.)

§1 Singularities of coherent sheaves and their cohomology classes

(1.1) First we recall the notions of profondeur and homological codimension (cf. [8], [10], and [22]). Let R be a Noetherian ring and M an R-module of finite type. If $\mathfrak{a} \subset R$ is an ideal, a sequence of elements $f_1, \ldots, f_q \in \mathfrak{a}$ is called a regular M-sequence if f_i is not a zero-divisor of

$$M/\Sigma_{j=1}^{i-1} f_j M,$$

where

$$\Sigma_{j=1}^{0} f_j M = 0.$$

The maximum length of regular M-sequences is denoted by $\mathrm{prof}_{\mathfrak{a}}$ M. If R is a local ring with maximal ideal $\mathfrak{m}$, then $\mathrm{prof}_{\mathfrak{m}}$ M is denoted by $\mathrm{codh}_R M$, called the homological codimension of M. The following lemma is proved in [10, p. 41, Cor. 3.5].

(1.2) Lemma. If $f \in \mathfrak{a}$ is not a zero-divisor of M, then

$$\mathrm{prof}_{\mathfrak{a}} M/fM = \mathrm{prof}_{\mathfrak{a}} M - 1.$$

(1.3) Lemma. Suppose S is another Noetherian ring and suppose there are ideals $\mathfrak{i}$, $\mathfrak{a}_S \subset S$ such that $R = S/\mathfrak{i}$ and $\mathfrak{a} = \mathfrak{a}_S/\mathfrak{i}$. If M is naturally considered as an S-module M_S, then

$$\mathrm{prof}_{\mathfrak{a}} M = \mathrm{prof}_{\mathfrak{a}_S} M.$$

The proof follows immediately from the fact that $f_1, \ldots, f_q \in \mathfrak{a}_S$ form a regular M_S-sequence if and only if their residue classes in $\mathfrak{a}$ form a regular M-sequence.

Since R is Noetherian and M is of finite type, there exist resolutions

$$\ldots \to R^{p_i} \to R^{p_{i-1}} \to \ldots \to R^{p_1} \to R^{p_0} \to M \to 0$$

of M. The minimal length of such resolutions is independent of the specific resolution (cf. [22, Chap. IV]) and is denoted by $dh_R M$, called the homological dimension of M over R. This number may be infinite. M is free if and only if $dh_R M = 0$.

(1.4) Lemma. If

$$0 \to K \to R^p \to M \to 0$$

is an exact sequence of R-modules, then

(a) K is free if M is free, and

(b) $dh_R K = dh_R M - 1$ if $dh_R M \geqq 1$ (see [22, IV-28]).

(1.5) Theorem (Syzygy theorem). If R is a regular local ring, then for any R-module M of finite type

$$dh_R M \leqq \dim R < \infty$$

and

$$dh_R M + codh_R M = \dim R,$$

where dim R is the Krull dimension of R (see [22, IV-35, Prop. 21] and [22, IV-39, Cor. 2]).

(1.6) Lemma. Let R be a regular local ring and Q its field of quotients. Then for an R-module M of finite type the following two conditions are equivalent.

(a) $M \approx R^k$ for some k (i.e. M is free).

(b) If

$$R^p \xrightarrow{\alpha} R^q \to M \to 0$$

is a representation of M by the matrix $\alpha = (a_{ij})$, then

$$\operatorname{rank}_{R/\mathfrak{m}} (\bar{a}_{ij}) = \operatorname{rank}_Q (a_{ij}),$$

where $\mathfrak{m} \subset R$ is the maximal ideal and $\bar{a}_{ij}$ is the residue class of a_{ij} in $R/\mathfrak{m}$.

Proof. (a) => (b). If $M = R^k$, then

$$\dim_{R/\mathfrak{m}} M \otimes_R R/\mathfrak{m} = \dim_Q M \otimes_R Q = k$$

and, since the tensor product is right exact, from the exact sequences

$$(*)\quad \begin{array}{ccccccc} (R/\mathfrak{m})^p & \to & (R/\mathfrak{m})^q & \to & M \otimes_R R/\mathfrak{m} & \to & 0 \\ Q^p & \to & Q^q & \to & M \otimes_R Q & \to & 0, \end{array}$$

we obtain (b).

If (b) is true, then by (*),

$$\dim_{R/\mathfrak{m}} M/\mathfrak{m} M = \dim_Q M \otimes_R Q$$

and we denote that number by k. By Krull-Nakayama lemma we can find an epimorphism

$$R^k \to M \to 0$$

and a representation

$$R^\ell \xrightarrow{\beta} R^k \to M \to 0.$$

By tensoring with $\otimes_R Q$, we obtain the isomorphism

$$Q^k \approx M \otimes_R Q.$$

Hence $\beta = 0$ and $M \approx R^k$. Q.E.D.

(1.7) Definitions. Let $(X, \mathcal{O})$ be a complex space and $\mathfrak{F}$ be a coherent analytic sheaf on X. For $x \in X$ we define

$$\operatorname{codh}_x \mathcal{F} = \begin{cases} \infty & \text{if } \mathcal{F}_x = 0 \\ \operatorname{codh}_{\mathcal{O}_x} \mathcal{F}_x & \text{if } \mathcal{F}_x \neq 0 \end{cases}$$

and define

$$\operatorname{codh} \mathcal{F} = \inf_{x \in X} \operatorname{codh}_x \mathcal{F} .$$

Further we define the <u>singularity subvarities</u> of $\mathcal{F}$ to be

$$S_m(\mathcal{F}) = \{x \in X \mid \operatorname{codh}_x \mathcal{F} \leqq m\}$$

for $m \geqq 0$. (It will be proved in (1.11) that $S_m(\mathcal{F})$ is a subvariety of X.) If $A \subset X$ is a subvariety and if $\mathcal{A} = \mathcal{I}(A)$ is its ideal-sheaf, then for $x \in A$ we define

$$\operatorname{prof}_{A,x} \mathcal{F} = \begin{cases} \infty & \text{if } \mathcal{F}_x = 0 \\ \operatorname{prof}_{\mathcal{A}_x} \mathcal{F}_x & \text{if } \mathcal{F}_x \neq 0 \end{cases}$$

and define

$$\operatorname{prof}_A \mathcal{F} = \inf_{x \in A} \operatorname{prof}_{A,x} \mathcal{F} .$$

(1.8) <u>Proposition</u>. The functions $x \mapsto \operatorname{prof}_{A,x} \mathcal{F}$ and $x \mapsto \operatorname{codh}_x \mathcal{F}$ are lower semi-continuous on A and X respectively.

Proof. (a) If $\text{prof}_{A,x}\mathcal{F} = \infty$, there is nothing to prove. So assume $\text{prof}_{A,x}\mathcal{F} = q < \infty$. Let $(f_1)_x,\ldots,(f_q)_x \in \mathcal{A}_x$ be a regular $\mathcal{F}_x$-sequence and let $f_1,\ldots,f_q \in \Gamma(U,\mathcal{A})$ be representing sections in some neighborhood U of x. Let $\mathcal{K}_i$ be the kernel of the homomorphism

$$\mathcal{F}/\Sigma_{j=1}^{i-1} f_j\mathcal{F} \rightarrow \mathcal{F}/\Sigma_{j=1}^{i-1} f_j\mathcal{F}$$

defined by multiplication by f_i on U ($1\leqq i\leqq q$). Since $(\mathcal{K}_i)_x = 0$ there exists a neighborhood V of x such that $\mathcal{K}_i|V = 0$. Hence, for $y \in V \cap A$, the germs $(f_1)_y,\ldots,(f_q)_y \in \mathcal{A}_y$ form a regular $\mathcal{F}_y$-sequence and so

$$\text{prof}_{A,y}\mathcal{F} \geqq q$$

for $y \in V \cap A$.

(b) The fact that $x \mapsto \text{codh}_x\mathcal{F}$ is lower semi-continuous on X is much deeper. Since semi-continuity is a local property, we can assume that X is a subvariety in some domain $D \subset \mathbb{C}^n$ and that

$$\mathcal{O} = (\mathcal{O}_D/\mathfrak{X})|X,$$

where $\mathcal{O}_D$ is the sheaf of germs of holomorphic functions on D and $\mathfrak{X} \subset \mathcal{O}_D$ is a coherent ideal-sheaf with $V(\mathfrak{X}) = X$. If we consider $\mathcal{F}$ as a coherent $\mathcal{O}_D$-sheaf, then by Lemma (1.3)

$$\mathrm{codh}_x \mathcal{F} = \mathrm{codh}_{(\mathcal{O}_D)_x} \mathcal{F}_x$$

and by Theorem (1.5)

$$\mathrm{codh}_x \mathcal{F} = n - \mathrm{dh}_{(\mathcal{O}_D)_x} \mathcal{F}_x .$$

So we need only prove that

$$x \mapsto \mathrm{dh}_{(\mathcal{O}_D)_x} \mathcal{F}_x$$

is upper semi-continuous. By coherence we can find a neighborhood V of x in D and a resolution of $\mathcal{F}$ on D:

$$0 \to \mathcal{K} \to \mathcal{O}_D^{p_{d-1}} \to \dots \to \mathcal{O}_D^{p_0} \to \mathcal{F} \to 0,$$

where

$$d = \mathrm{dh}_{(\mathcal{O}_D)_x} \mathcal{F}_x .$$

By Lemma (1.4),

$$\mathrm{dh}_{(\mathcal{O}_D)_x} \mathcal{K}_x = 0 .$$

Hence $\mathcal{K}_x$ is free and there exists a neighborhood U of x in V such that $\mathcal{K} \approx \mathcal{O}_D^p$ on U. So, for any $y \in U$,

$$\mathrm{dh}_{(\mathcal{O}_D)_y} \mathcal{F}_y \leq d .$$

Q.E.D.

(1.9) <u>Proposition</u>. Let A be a subvariety in a complex space $(X, \mathcal{O})$ and $\mathcal{F}$ be a coherent analytic sheaf on X. Then the sheaf $\mathcal{H}^0_A \mathcal{F}$ is coherent and is equal to the subsheaf of all section germs of $\mathcal{F}$ annihilated by some power of $\mathcal{A} = \mathcal{I}(A)$.

Proof. Let $\mathcal{F}_k$ denote the kernel of the sheaf-homomorphism

$$\mathcal{F} \mapsto \mathcal{H}om_{\mathcal{O}}(\mathcal{A}^k, \mathcal{F})$$

defined by

$$s \mapsto (f \mapsto fs).$$

We have the ascending chain

$$\mathcal{F}_1 \subset \mathcal{F}_2 \subset \ldots \subset \mathcal{F}$$

By (0.12)

$$\mathcal{F}_\infty := \cup_{k=1}^{\infty} \mathcal{F}_k$$

is locally the same as some $\mathcal{F}_k$ and hence is coherent. The proposition will follow if we can prove that $\mathcal{F}_\infty = \mathcal{H}^0_A \mathcal{F}$. It is clear that $\mathcal{F}_\infty \subset \mathcal{H}^0_A \mathcal{F}$. To prove the other direction, we take

$s_x \in (\mathcal{H}^0_A \mathcal{F})_x$ and let $s \in \Gamma_A(U, \mathcal{F})$ be a representing section in some open neighborhood U of x. Let $\mathcal{B}$ be the annihilator sheaf of s, i.e. $\mathcal{B}$ is the maximum ideal-sheaf on U satisfying $\mathcal{B}s = 0$. $\mathcal{B}$ is coherent, because it is the kernel of the sheaf-homomorphism $\mathcal{O} \to \mathcal{F}$ defined by $f \mapsto fs$. Since Supp $s \subset A$, we have $V(\mathcal{B}) \subset A \cap U$. By the Hilbert Nullstellensatz (0.10), after shrinking U, we have $\mathcal{A}^k|U \subset \mathcal{B}$. Hence $\mathcal{A}^k s = 0$ on U and $s_x \in (\mathcal{F}_\infty)_x$. Q.E.D.

(1.10) <u>Proposition</u>. Let A be a subvariety of a complex space $(X, \mathcal{O})$ and $\mathcal{F}$ be a coherent analytic sheaf on X. For a point $x \in A$ the following two conditions are equivalent.

(a) $(\mathcal{H}^0_A \mathcal{F})_x = 0$.

(b) $\operatorname{prof}_{A,x} \mathcal{F} \geqq 1$.

Proof. Let $\mathcal{A}$ be the coherent ideal-sheaf of A.

If (b) is satisfied, then there exists an element $f_x \in \mathcal{A}_x$ which is not a zero-divisor for $\mathcal{F}_x$. If $s_x \in (\mathcal{H}^0_A \mathcal{F})_x$, then by (0.9) we have $\mathcal{A}^k_x s_x = 0$ in $\mathcal{F}_x$ for some k. In particular, $f^k_x s_x = 0$. But then $s_x = 0$ since f_x is not a zero-divisor. So $(\mathcal{H}^0_A \mathcal{F})_x = 0$.

Assume conversely that $(\mathcal{H}^0_A \mathcal{F})_x = 0$ and $\operatorname{prof}_{\mathcal{A}_x} \mathcal{F}_x = 0$. Then every element $f_x \in \mathcal{A}_x$ is a zero-divisor of $\mathcal{F}_x$ and so $\mathcal{A}_x$ is contained in some associated prime ideal $\mathcal{J} \subset \mathcal{O}_x$ of $\mathcal{F}_x$. Since $\mathcal{J}$ is the annihilator of some $s_x \neq 0$ in $\mathcal{F}_x$, we have $\mathcal{A}_x s_x = 0$ and hence $\mathcal{A}s = 0$ for some representing section s in some open neighborborhood U of x. Hence Supp s $\subset$ A $\cap$ U and $s_x \in (\mathcal{H}^0_A \mathcal{F})_x = 0$. Contradiction. Q.E.D.

(1.11) Theorem. Let $(X, \mathcal{O})$ be a complex space and $\mathcal{F}$ be a coherent analytic sheaf on X. Then the sets $S_m(\mathcal{F})$ are subvarieties of X and $\dim S_m(\mathcal{F}) \leq m$.

Proof. (a) By Proposition (1.8) the sets $S_m(\mathcal{F})$ are closed. So the proof is only local in nature. As in the proof of Proposition (1.8), by embedding X locally as a subvariety of a domain D in $\mathbb{C}^n$, we need only consider the special case X = D.

We first prove that $S_{n-1}(\mathcal{F})$ is a subvariety. Let $x \in S_{n-1}(\mathcal{F})$ and let U be an open neighborhood of x such that there exists a resolution

$$\mathcal{O}^p_D \xrightarrow{\alpha} \mathcal{O}^q_D \rightarrow \mathcal{F} \rightarrow 0$$

on U, where $\mathcal{O}_D$ is the sheaf of germs of holomorphic functions on D. α is defined by a matrix (a_{ij}) of holomorphic functions on U. By Lemma (1.6) for $y \in U$ the module $\mathcal{F}_y$ is free if and only if

$$\operatorname{rank}_{(\mathcal{O}_D)_y/\mathfrak{m}_y}((a_{ij})_y + \mathfrak{m}_y) = \operatorname{rank}((a_{ij})_y)$$

where $\mathfrak{m}_y$ is the maximal ideal of $(\mathcal{O}_D)_y$ and $(a_{ij})_y + \mathfrak{m}_y$ is the residue class of $(a_{ij})_y$ in $(\mathcal{O}_D)_y/\mathfrak{m}_y$ (which is the same as the value $a_{ij}(y)$ of the function a_{ij} at y when $(\mathcal{O}_D)_y/\mathfrak{m}_y$ is identified canonically with $\mathbb{C}$). If U is connected (which we may assume), then $\mathrm{rank}((a_{ij})_y)$ is a constant r independent of y. Since $\mathcal{F}_y$ is free if and only if $\mathrm{codh}_y \mathcal{F} = n$, by Theorem (1.5) we have

$$S_{n-1}(\mathcal{F}) \cap U = \{y \in U \mid \mathrm{rank}_{\mathbb{C}} (a_{ij}(y)) < r\}.$$

Hence $S_{n-1}(\mathcal{F})$ is the zero-set of subdeterminants of (a_{ij}) of rank r and so is a subvariety.

For the general case, let $\mathcal{K} = \mathrm{Im}\ \alpha$. From the exact sequence

$$0 \to \mathcal{K} \to \mathcal{O}_D^q \to \mathcal{F} \to 0$$

on U and by Lemma (1.4) and Theorem (1.5) we obtain

$$S_m(\mathcal{F}) \cap U = S_{m+1}(\mathcal{K}) \quad \text{for } m \leqq n - 2.$$

So by induction on m all $S_m(\mathcal{F})$ are subvarieties.

(b) To prove the dimension estimates, we use Propositions (1.8), (1.9), and (1.10). First we prove that $S_0(\mathcal{F})$ is at most a discrete set. Let $x \in X$ and

$$\mathcal{G} = \mathcal{F}/\mathcal{H}^0_{\{x\}}\mathcal{F}$$

which is coherent by Proposition (1.9). We have $\mathcal{H}^0_{\{x\}}\mathcal{G} = 0$ and hence by Proposition (1.10) $\mathrm{codh}_x \mathcal{G} \geqq 1$. By Proposition (1.8) there is an open neighborhood U of x such that $\mathrm{codh}_y \mathcal{G} \geqq 1$ for

$y \in U$. Since for $y \neq x$ we have $\mathcal{G}_y = \mathcal{F}_y$, it follows that $S_0(\mathcal{F}) \cap U$ is either $\{x\}$ or empty.

If $m \geqq 1$, assume that $\dim S_k(\mathcal{F}) \leqq k$ for $k < m$. Let $x \in S_m(\mathcal{F})$. Without loss of generality we may assume that $x \notin S_0(\mathcal{F})$, for, if $\dim_x S_m(\mathcal{F}) > m$, then we can find $y \neq x$ in a neighborhood of x such that $\dim_y S_m(\mathcal{F}) > m$. Since $x \notin S_0(\mathcal{F})$, there exists $f_x \in \mathfrak{m}_x$ such that f_x is not a zero-divisor for $\mathcal{F}_x$. On some open neighborhood U of x we can find a representing function $f \in \Gamma(U, \mathcal{O}_D)$ for f_x such that the sheaf-homomorphism $\mathcal{F} \to \mathcal{F}$ on U defined by multiplication by f is injective. Let

$$F = \{y \in U | f(y) = 0\}.$$

By (0.17)

$$\dim_x S_m(\mathcal{F}) - 1 \leqq \dim_x F \cap S_m(\mathcal{F}).$$

However, from Lemma (1.2) we conclude that

$$F \cap S_m(\mathcal{F}) \cap U \subset S_{m-1}(\mathcal{F}/f\mathcal{F}) \cap U$$

and by induction hypothesis we know that

$$\dim_x S_{m-1}(\mathcal{F}/f\mathcal{F}) \leqq m - 1 .$$

Hence $\dim S_m(\mathcal{F}) \leqq m$. Q.E.D.

(1.12) Proposition. Let A be a subvariety of dimension d in a domain D in $\mathbb{C}^n$ and $\mathcal{O}$ be the sheaf of germs of holomorphic functions on D. Then $\mathcal{H}^i_A \mathcal{O} = 0$ for $i < n - d$.

Proof. First let $x \in A$ be a regular point. By a biholomorphic transformation of an open neighborhood of x we may assume that $x = 0$ and

$$A \cap U = \{z_{d+1} = \dots = z_n = 0\}$$

for some open neighborhood U of x. For any polydisc neighborhood Δ of x in U by (0.14)(F_2) we have

$$\begin{cases} H^0(\Delta, \mathcal{O}) = H^0(\Delta - A, \mathcal{O}) \\ H^i(\Delta - A, \mathcal{O}) = 0 \quad \text{for} \quad 1 \leq i < n - d - 1. \end{cases}$$

Hence by (0.3) and Cartan's theorem B

$$H^i_A(\Delta, \mathcal{O}) = 0 \quad \text{for } i < n - d$$

and $(\mathcal{H}^i_A \mathcal{O})_x = 0$ for $i < n - d$.

If $x \in A$ is not regular, we proceed by induction on d. For $d = 0$ the proposition is already proved. Let A' be the subvariety of all singular points of A and let $A'' = A - A'$. We have the exact sequence

$$\dots \to \mathcal{H}^i_{A'} \mathcal{O} \to \mathcal{H}^i_A \mathcal{O} \to \mathcal{H}^i_{A''} \mathcal{O} \to \dots \quad .$$

Since dim A' $\leqq$ d - 1, by induction hypothesis we have $\mathcal{H}^i_{A'}\mathcal{O} = 0$ for i < n - d. Let D' = D - A'. By the first part of the proof, $\mathcal{H}^i_{A''}\mathcal{O} = 0$ on D' for i < n - d. By the excision theorem and Lemma (0.6), for any open subset U of D,

$$H^i_{A''}(U, \mathcal{O}) = H^i_{A''}(D' \cap U, \mathcal{O}) = \Gamma(D' \cap U, \mathcal{H}^i_{A''}\mathcal{O}) = 0 .$$

Hence $\mathcal{H}^i_{A''}\mathcal{O} = 0$ for i < n - d. It follows that $\mathcal{H}^i_A\mathcal{O} = 0$ for i < n - d. Q.E.D.

(1.13) Proposition. If A is a subvariety of dimension d in a domain D in $\mathbb{C}^n$ and $\mathcal{F}$ is a coherent analytic sheaf on D such that codh $\mathcal{F} \geqq d + q$, then $\mathcal{H}^i_A\mathcal{F} = 0$ for $0 \leqq i < q$.

Proof. We proceed by descending induction on codh $\mathcal{F}$. If $\mathcal{F}$ is locally free, then $q \leqq n - d$ and by (1.12) we have the result. If $\mathcal{F}_x$ is not free, we have n > d - q and from a representation of $\mathcal{F}$ in an open neighborhood U of x

$$(*) \qquad 0 \to \mathcal{K} \to \mathcal{O}^s|U \to \mathcal{F}|U \to 0$$

(where $\mathcal{O}$ is the sheaf of germs of holomorphic function on D) we obtain

$$\text{codh}\, \mathcal{K} \geqq \text{codh}\, \mathcal{F}|U + 1.$$

By induction hypothesis, $\mathcal{H}^i_A\mathcal{K} = 0$ for $0 \leqq i < q + 1$. Since

$n \geqq d + q + 1$, we know from (1.12) that $\mathcal{H}^i_A \mathcal{O} = 0$ for $i < q + 1$. Hence from (*) we obtain the isomorphisms $\mathcal{H}^i_A \mathcal{F} \approx \mathcal{H}^{i+1}_A \mathcal{K} = 0$ for $i < q$. Q.E.D.

Proposition (1.13) remains true if we assume only that A is a subvariety of dimension d in an open subset D' of D. For, $\mathcal{H}^i_A \mathcal{F} = 0$ on D' for $i < q$ and by the excision theorem and Lemma (0.6), for any open subset U of D,

$$H^i_A(U, \mathcal{F}) = H^i_A(D' \cap U, \mathcal{F}) = \Gamma(D' \cap U, \mathcal{H}^i_A \mathcal{F}) = 0.$$

(1.14) <u>Theorem.</u> Let $(X, \mathcal{O})$ be a complex space, $A \subset X$ a subvariety, and $\mathcal{F}$ a coherent analytic sheaf on X. Then for any integer $q \geqq 0$ the following four conditions are equivalent.

(a) $\mathrm{prof}_A \mathcal{F} \geqq q + 1$.

(b) $\dim A \cap S_{k+q+1}(\mathcal{F}) \leqq k$ for all k.

(c) $\mathcal{H}^i_A \mathcal{F} = 0$ for $i \leqq q$.

(d) For any open subset U of X the restriction maps

$$H^i(U, \mathcal{F}) \to H^i(U-A, \mathcal{F})$$

are bijective for $i < q$ and injective for $i = q$.

Proof. The equivalence (c) <=> (d) follows directly from Corollary (0.7).

(a) => (b). Let $\mathcal{A}$ be the ideal-sheaf of A and let $\mathrm{prof}_{\mathcal{A}_x} \mathcal{F}_x \geq q + 1$. There exists an open neighborhood U of x and a sequence $f_1,\dots,f_{q+1} \in \Gamma(U,\mathcal{A})$ such that for each $y \in U \cap A$ the germs $(f_i)_y$ $(1 \leq i \leq q+1)$ form a regular $\mathcal{F}_y$-sequence. By applying Lemma (1.2) q + 1 times we obtain

$$A \cap U \cap S_{k+q+1}(\mathcal{F}) = A \cap U \cap S_k(\mathcal{F}/\Sigma_{i=1}^{q+1} f_i \mathcal{F})$$

and by Theorem (1.11)

$$\dim_x A \cap S_{k+q+1}(\mathcal{F}) \leq k.$$

(b) => (c) is proved by induction on dim A. Let $x \in A$ and $\dim_x A = 0$. Then

$$\dim A \cap S_q(\mathcal{F}) \leq -1$$

or

$$S_q(\mathcal{F}) \cap U = \emptyset$$

for some open neighborhood U of x and

$$\mathrm{codh}\ \mathcal{F}|U \geq q + 1.$$

By Proposition (1.13) we get $\mathcal{H}_A^i \mathcal{F}|U = 0$ for $i \leq q$. Now let $k = d - 1$ and assume $\dim_x A = d$. Then

$$A' := A \cap S_{q+d}(\mathcal{F})$$

has dimension $\leqq$ d - 1. Let A" = A - A'. By Proposition (1.13) we have $\mathcal{H}^i_{A''}\mathcal{F} = 0$ for $i \leqq q$. From the canonical sequence

$$\cdots \to \mathcal{H}^i_{A'}\mathcal{F} \to \mathcal{H}^i_A\mathcal{F} \to \mathcal{H}^i_{A''}\mathcal{F} \to \cdots$$

we obtain the result, because A' also satisfies (b).

(c) => (a). We are going to prove the following sharper result: if, for $x \in A$, $(\mathcal{H}^i_A\mathcal{F})_x = 0$ for $i \leqq q$, then $\mathrm{prof}_{A,x}\mathcal{F} \geqq q + 1$. If $q = 0$, Proposition (1.10) implies our assertion. If $q \geqq 1$, by Proposition (1.10) we can find an open neighborhood U of x and an $f \in \Gamma(U, \mathcal{A})$ such that the sequence

$$0 \to \mathcal{F}|U \overset{\alpha}{\to} \mathcal{F}|U \to \mathcal{F}/f\mathcal{F}|U \to 0$$

is exact, where α is defined by multiplication by f. From the sequence

$$\cdots \to \mathcal{H}^i_A\mathcal{F} \to \mathcal{H}^i_A(\mathcal{F}/f\mathcal{F}) \to \mathcal{H}^{i+1}_A\mathcal{F} \to \cdots$$

we obtain $\mathcal{H}^i_A(\mathcal{F}/f\mathcal{F})_x = 0$ for $i \leqq q - 1$. If we proceed by induction on q, we get

$$\mathrm{prof}_{\mathcal{A}_x}\mathcal{F}_x/f_x\mathcal{F}_x \geqq q$$

and by Lemma (1.2)

$$\mathrm{prof}_{\mathcal{A}_x}\mathcal{F}_x \geqq q + 1.$$

Q.E.D.

(1.15) <u>Corollary</u>. If, for $x \in A$, $(\mathcal{H}_A^i \mathcal{F})_x = 0$ for $i \leqq q$, then there exists an open neighborhood U of x such that $\mathcal{H}_A^i \mathcal{F} | U = 0$ for $i \leqq q$.

(1.16) <u>Corollary</u>. Let X be a complex space and $\mathcal{F}$ be a coherent analytic sheaf on X. Then the following two conditions are equivalent.

(a) $\dim S_{k+q+1}(\mathcal{F}) \leqq k$ for every $k < d$.

(b) For any subvariety A of dimension $\leqq d$ in an open subset U of X, $\mathcal{H}_A^i \mathcal{F} = 0$ for $i \leqq q$.

Proof. (a) => (b) follows directly from Theorem (1.14). (b) => (a) is proved in the following way. Assume (a) is not true for some $k < d$. Then we can find a subvariety A in $S_{k+q+1}(\mathcal{F}) \cap U$, where U is an open subset of X and $\dim A = k + 1 \leqq d$. By (b) and Theorem (1.14), on U we get

$$k + 1 = \dim A = \dim A \cap S_{k+q+1}(\mathcal{F}) \leqq k$$

which is a contradiction. Q.E.D.

(1.17) <u>Corollary</u>. Let $(X, \mathcal{O})$ be a complex space and $\mathcal{F}$ be a coherent analytic sheaf on X. Then the following four conditions are equivalent.

(a) $\dim S_{k+1}(\mathcal{F}) \leqq k$ for all $k < d$.

(b) $\mathcal{H}_A^0 \mathcal{F} = 0$ for any subvariety A of dimension $\leqq d$ in an open subset U of X.

(c) $\mathcal{F}$ contains no non-zero sections whose supports have dimension $\leqq$ d.

(d) For every $x \in X$ the $\mathcal{O}_x$-module $\mathcal{F}_x$ has no associated prime ideal of dimension $\leqq$ d.

Proof. All implications are trivial with the exception of (d) => (c). If $s_x \in \mathcal{F}_x$ is the germ of a local section s, then the annihilator sheaf $\mathcal{A}$ of s is coherent and $\mathcal{A}_x$ is contained in some associated prime ideal of $\mathcal{F}_x$. Since Supp s = $V(\mathcal{A})$, this means $\dim_x$ Supp s $\geqq$ d+1.

Q.E.D.

(1.18) Corollary. Suppose $(X, \mathcal{O})$ is a complex space. If $f \in \Gamma(X, \mathcal{O})$ and $V(f) = \mathrm{Supp}(\mathcal{O}/f\mathcal{O})$ is its subvariety, then for a coherent analytic sheaf $\mathcal{F}$ on X the following two conditions are equivalent.

(a) $\dim V(f) \cap S_{k+1}(\mathcal{F}) \leqq k$ for all k.

(b) f_x is not a zero-divisor of $\mathcal{F}_x$ for every $x \in X$.

(1.19) Remark. Let

$$p = (f,\varphi): (X, \mathcal{O}) \to (\mathbb{C}^n, {}_n\mathcal{O})$$

be a morphism of complex spaces and $\mathcal{F}$ be a coherent analytic sheaf on X. For every $x \in X$, φ defines a homomorphism

$$\varphi_x: {}_n\mathcal{O}_{f(x)} \to \mathcal{O}_x$$

which makes $\mathcal{F}_x$ an ${}_n\mathcal{O}_{f(x)}$-module. $\mathcal{F}$ is called p-flat at x if $\mathcal{F}_x$ is a flat ${}_n\mathcal{O}_{f(x)}$-module. Let $f = (f_1, \ldots, f_n)$ where $f_i \in \Gamma(X, \mathcal{O})$. It can be shown that $\mathcal{F}$ is p-flat if and only if $(f_i)_x - f_i(x)(1 \leqq i \leqq n)$ form a regular $\mathcal{F}_x$-sequence. Since the subvariety defined by $f_i - f_i(x)$ is the fiber $p^{-1}p(x)$ of p through x, we have the following: $\mathcal{F}$ is p-flat on X if and only if $\dim p^{-1}(t) \cap S_{k+n}(\mathcal{F}) \leqq k$ for every $t \in \mathbb{C}^n$ and every k.

Let D be a domain in $\mathbb{C}^n$ and $\mathcal{F}$ be a coherent analytic sheaf on D. $\mathcal{F}$ is called a kth syzygy-sheaf if there is an exact sequence

$$0 \to \mathcal{F} \to \mathcal{O}^{p_1} \to \ldots \to \mathcal{O}^{p_k}$$

on D, where $\mathcal{O}$ is the sheaf of germs of holomorphic functions on D.

(1.20) Proposition. For a relatively compact domain X in D whose closure X^- is holomorphically convex in D, the following two conditions are equivalent.

(a) $\dim X \cap S_m(\mathcal{F}) \leqq m - k$ for all $m < n$.

(b) $\mathcal{F}|X$ is a kth syzygy-sheaf on X.

Proof. (b) => (a) follows directly from Lemma (1.4) and Theorems (1.5) and (1.11). (a) => (b) is proved by induction on k as follows. Assume $k \geqq 1$. Let $\mathcal{K}$ be the kernel of the canonical homomorphism

$$\mathcal{F} \to \mathcal{F}^{**} := \mathcal{H}om_{\mathcal{O}}(\mathcal{H}om_{\mathcal{O}}(\mathcal{F}, \mathcal{O}), \mathcal{O}).$$

Since for $x \notin S_{n-1}(\mathcal{F})$ the module $\mathcal{F}_x$ is free, we have $\mathcal{K}_x = 0$. Hence

$$\operatorname{Supp} \mathcal{K} \subset S_{n-1}(\mathcal{F}).$$

Since

$$\dim X \cap S_{n-1}(\mathcal{F}) \leqq n - 2,$$

by Corollary (1.17) we have

$$\mathcal{K}|X = \mathcal{H}^0_{S_{n-1}(\mathcal{F})}\mathcal{K}|X = 0.$$

So

$$\mathcal{F}|X \subset \mathcal{F}^{**}|X.$$

Since X^- is compact and holomorphically convex, by Cartan's theorem A we can find an epimorphism

$$\mathcal{O}^p \to \mathcal{F}^* \to 0$$

on an open neighborhood of X^-. Hence $\mathcal{F}^{**} \subset \mathcal{O}^p$ in that open neighborhood and we get a homomorphism $\alpha: \mathcal{F} \to \mathcal{O}^p$ which is injective on X. Let $\mathcal{G} = \mathcal{O}^p/\alpha(\mathcal{F})$. Then by Lemma (1.4) and Theorem (1.5)

$$S_m(\mathcal{G}) \cap X = S_{m+1}(\mathcal{F}) \cap X \text{ for } m < n.$$

Hence

$$\dim X \cap S_m(\mathcal{G}) \leqq m - (k-1) \text{ for } m < n .$$

By induction hypothesis $\mathcal{G}|X$ is a $(k-1)^{th}$ syzygy-sheaf and the proposition is proved. Q.E.D.

(1.21) <u>Corollary</u>. $\mathcal{F}$ is reflexive (i.e. $\mathcal{F} = \mathcal{F}^{**}$) if and only if $\dim S_m(\mathcal{F}) \leqq m - 2$ on D for all $m < n$.

Proof. If $\mathcal{F} = \mathcal{F}^{**}$, then by a local resolution

$$\mathcal{O}^{p_2} \to \mathcal{O}^{p_1} \to \mathcal{F}^* \to 0$$

we conclude that $\mathcal{F}$ is locally a 2nd syzygy-sheaf and therefore $\dim S_m(\mathcal{F}) \leqq m - 2$ on D for all $m < n$.

If conversely $\dim S_m(\mathcal{F}) \leqq m - 2$ on D for all $m < n$, by Corollary (1.17)

$$\mathcal{H}^0_A\mathcal{F} = \mathcal{H}^1_A\mathcal{F} = 0$$

for any subvariety A of dimension $\leqq n - 2$ in D. Let $S = S_{n-1}(\mathcal{F})$. is a subvariety of dimension $\leqq n - 3$ in D. The kernel of the natural sheaf-homomorphism α: $\mathcal{F} \to \mathcal{F}^{**}$ is contained in $\mathcal{H}^0_S\mathcal{F}$ and hence is 0. Consider the exact sequence

$$0 \to \mathcal{F} \overset{\alpha}{\to} \mathcal{F}^{**} \to \mathcal{L} \to 0 ,$$

where $\mathcal{L}$ = Coker α. Since Supp $\mathcal{L} \subset S$, from the exact sequence

$$\ldots \to \mathcal{H}^0_S \mathcal{F}^{**} \to \mathcal{H}^0_S \mathcal{L} \to \mathcal{H}^1_S \mathcal{F} \to \ldots$$

we conclude that $\mathcal{L} = \mathcal{H}^0_S \mathcal{L} = 0$. Q.E.D.

§2 Primary decomposition and relative gap-sheaves

Let $(X, \mathcal{O})$ be a complex space and $\mathcal{F}$ be a coherent analytic sheaf on X. Then the zero-submodule $0_x \subset \mathcal{F}_x$ admits a primary decomposition $0_x = \bigcap_{\rho=1}^{r} Q_\rho$ with associated prime ideals $\mathfrak{p}_\rho \subset \mathcal{O}_x$. If we choose local sections in $\mathcal{F}$ and $\mathcal{O}$ which represent the generators of Q_ρ and $\mathfrak{p}_\rho$, we can define on an open neighborhood U of x coherent analytic subsheaves $\mathcal{Q}_\rho \subset \mathcal{F}|U$ and $\mathcal{P}_\rho \subset \mathcal{O}|U$ such that $(\mathcal{Q}_\rho)_x = Q_\rho$ and $(\mathcal{P}_\rho)_x = \mathfrak{p}_\rho$. The ideal-sheaves $\mathcal{P}_\rho$ define subvarieties $P_\rho \subset U$ such that the germs $(P_\rho)_x$ are irreducible. We call the $(P_\rho)_x$ <u>subvariety germs associated to</u> $\mathcal{F}$ <u>at</u> x and denote them by $P_{\rho x}(\mathcal{F})$. Since the $\mathfrak{p}_\rho$ are uniquely determined, the $P_{\rho x}(\mathcal{F})$ are uniquely determined by $\mathcal{F}_x$.

(2.1) <u>Lemma</u>. If A is a subvariety of X, then for any point x there exists an open neighborhood U of x such that

$$\mathcal{H}_A^0 \mathcal{F}|U = \bigcap \{\mathcal{Q}_\rho \mid (P_\rho)_x \not\subset A_x\} ,$$

where the $\mathcal{Q}_\rho$ define a primary decomposition of $\mathcal{F}_x$ at x.

Proof. If $(P_\rho)_x \not\subset A_x$, then $\mathcal{A}_x \not\subset (\mathcal{P}_\rho)_x$, where $\mathcal{A}$ is the ideal-sheaf of A. Hence there exists an $f_x \in \mathcal{A}_x$ which is not a zero-divisor of $\mathcal{F}_x/(\mathcal{Q}_\rho)_x$. By Proposition (1.10) we have $\mathcal{H}_A^0(\mathcal{F}/\mathcal{Q}_\rho) = 0$ in some open neighborhood of x and from the exact sequence

$$0 \to \mathcal{H}_A^0 \mathcal{Q}_\rho \to \mathcal{H}_A^0 \mathcal{F} \to \mathcal{H}_A^0(\mathcal{F}/\mathcal{Q}_\rho)$$

we obtain

$$\mathcal{H}_A^0 \mathcal{F} = \mathcal{H}_A^0 \mathcal{Q}_\rho \subset \mathcal{Q}_\rho .$$

Hence there exists an open neighborhood U of x such that

$$\mathcal{H}_A^0 \mathcal{F} | U \subset \cap \{\mathcal{Q}_\rho | (P_\rho)_x \not\subset A_x\} .$$

Since for each ρ

$$(\mathcal{P}_\rho)_x^{k_\rho} \mathcal{F}_x \subset (\mathcal{Q}_\rho)_x$$

for some k_ρ, for any

$$s_x \in \cap \{(\mathcal{Q}_\rho)_x \mid (P_\rho)_x \not\subset A_x\}$$

we have

$$\Pi\{(\mathcal{P}_\sigma)_x^{k_\sigma} s_x \mid (P_\sigma)_x \subset A_x\} = 0,$$

because the elements in this set are contained in $\cap_{\rho=1}^r (\mathcal{Q}_\rho)_x = 0$. Hence $\mathcal{A}_x^m s_x = 0$ for some m. It follows that $s_x \in (\mathcal{H}_A^0 \mathcal{F})_x$. Q.E.D.

Let $\mathcal{G}$ be another coherent analytic sheaf on X and assume $\mathcal{F} \subset \mathcal{G}$. For a subvariety A of X we define the subsheaf $\mathcal{F}[A]$ of $\mathcal{G}$ by the following presheaf:

$$U \mapsto \{s \in \Gamma(U, \mathcal{G}) \mid s|U-A \in \Gamma(U-A, \mathcal{F})\} .$$

We clearly have

$$\mathcal{F} \subset \mathcal{F}[A] \subset \mathcal{G}$$

and

$$(\mathcal{F}[A])[A] = \mathcal{F}[A].$$

The sheaf $\mathcal{F}[A]$ is called the <u>relative gap-sheaf of</u> $\mathcal{F}$ <u>in</u> $\mathcal{G}$ <u>with</u> <u>respect to</u> A.

If we look at the canonical exact sequence

$$0 \to \mathcal{F} \to \mathcal{G} \overset{\pi}{\to} \mathcal{G}/\mathcal{F} \to 0$$

with projection π, then we have

$$\mathcal{F}[A] = \pi^{-1} \mathcal{H}^0_A(\mathcal{G}/\mathcal{F}).$$

By Proposition (1.9) and Lemma (2.1) we get at once the following theorem.

(2.2) <u>Theorem</u>. If $\mathcal{F} \subset \mathcal{G}$ are coherent analytic sheaves on a complex space X and A is a subvariety, then the sheaf $\mathcal{F}[A]$ is coherent. A section germ $s_x \in \mathcal{G}_x$ is in $\mathcal{F}[A]_x$ if and only if $\mathcal{A}_x^k s_x \subset \mathcal{F}_x$ for some integer k, where $\mathcal{A}$ is the ideal-sheaf of A. Moreover, if $\mathcal{F}_x = \cap_{\rho=1}^r (\mathcal{Q}_\rho)_x$ is a primary decomposition of $\mathcal{F}_x$ in $\mathcal{G}_x$ and $(P_\rho)_x$ $(1 \leq \rho \leq r)$ are the subvariety germs at x defined by the associated prime ideals, then

$$\mathcal{J}[A] = \cap \{\mathcal{Q}_\rho \mid (P_\rho)_x \not\subset A_x\}$$

in some open neighborhood of x, where $\mathcal{Q}_\rho$ is a coherent analytic subsheaf of $\mathcal{G}$ on an open neighborhood of x whose stalk at x is $(\mathcal{Q}_\rho)_x$.

For the proof we need only observe that $s_x \in \mathcal{J}[A]_x$ if and only if $\pi_x s_x \in \mathcal{H}^0_A(\mathcal{G}/\mathcal{J})_x$ which is equivalent to

$$\pi_x \mathcal{A}^k_x s_x = \mathcal{A}^k_x \pi_x s_x = 0$$

for some k. The primary decomposition of $\mathcal{J}_x$ in $\mathcal{G}_x$ is equivalent to the primary decomposition of the zero-submodule of $(\mathcal{G}/\mathcal{J})_x$ and $(P_\rho)_x = P_{\rho x}(\mathcal{G}/\mathcal{J})$. Hence the last assertion follows from Lemma (2.1).

Let $d \geqq 0$ be an integer and $U \subset X$ an open subset. By $\mathfrak{A}_d(U)$ we denote the set of all subvarieties of dimension $\leqq d$ in U. Let $\mathfrak{A}_d = \cup \mathfrak{A}_d(U)$ where the union is over all open subsets U of X. $\mathfrak{A}_d$ and each $\mathfrak{A}_d(U)$ are directed sets in the ordering defined by inclusion. If $A \subset B$ are in $\mathfrak{A}_d$, we have a natural sheaf-homomorphism $\mathcal{H}^0_A \mathcal{J} \to \mathcal{H}^0_B \mathcal{J}$ for any coherent analytic sheaf $\mathcal{J}$ on X. (Note that $A \in \mathfrak{A}_d$ is a locally closed subset which is analytic at each of its points). We define

$$\mathcal{H}^0_d \mathcal{J} = \varinjlim \{\mathcal{H}^0_A \mathcal{J} \mid A \in \mathfrak{A}_d\} .$$

This sheaf is the subsheaf of $\mathcal{J}$ consisting of all section germs whose supports have dimension $\leq$ d.

(2.3) <u>Lemma</u>. For any coherent analytic sheaf $\mathcal{F}$ on X and any $d \geq 0$ the sheaf $\mathcal{H}^0_d \mathcal{F}$ is coherent and is equal to $\mathcal{H}^0_{S_d(\mathcal{F})} \mathcal{F}$. Moreover, the d-dimensional branches of Supp $\mathcal{H}^0_d \mathcal{F}$ and $S_d(\mathcal{F})$ are the same.

Proof. Since dim $S_d(\mathcal{F}) \leq d$, we have $\mathcal{H}^0_{S_d(\mathcal{F})} \mathcal{F} \subset \mathcal{H}^0_d \mathcal{F}$. If $x \notin S_d(\mathcal{F})$, we can find an open neighborhood U of x such that $U \cap S_d(\mathcal{F}) = \emptyset$. By Corollary (1.17) $\mathcal{H}^0_A \mathcal{F} | U = 0$ for any $A \in \mathfrak{A}_d(U)$. Hence Supp $\mathcal{H}^0_d \mathcal{F} \subset S_d(\mathcal{F})$. $\mathcal{H}^0_d \mathcal{F}$ agrees with $\mathcal{H}^0_{S_d(\mathcal{F})} \mathcal{F}$ and is therefore coherent.

For the last statement we need only prove that any d-dimensional branch S of $S_d(\mathcal{F})$ is also a branch of Supp $\mathcal{H}^0_d \mathcal{F}$. Since dim $S_{d-1}(\mathcal{F}) \leq d - 1$, there exists $x \in S - S_{d-1}(\mathcal{F})$. By Corollary (1.17) there exist a neighborhood U of x disjoint from $S_{d-1}(\mathcal{F})$ and and $A \in \mathfrak{A}_d(U)$ containing x such that $\mathcal{H}^0_A \mathcal{F} | U \neq 0$. Hence $x \in$ Supp $\mathcal{H}^0_d \mathcal{F}$ and

$$S - S_{d-1}(\mathcal{F}) \subset \text{Supp } \mathcal{H}^0_d \mathcal{F} .$$

Since the support of any coherent analytic sheaf is closed, we obtain $S \subset$ Supp $\mathcal{H}^0_d \mathcal{F}$. Q.E.D.

(2.4) <u>Corollary</u>. Supp$(\mathcal{H}^0_d \mathcal{F} / \mathcal{H}^0_{d-1} \mathcal{F})$ is the union of all d-dimensional branches of $S_d(\mathcal{F})$.

(2.5) Lemma. If $\mathcal{F}$ is a coherent analytic sheaf on a complex space X and $d \geqq 0$, then for any $x \in X$ there is an open neighborhood U of x for which

$$\mathcal{H}_d^0 \mathcal{F} | U = \cap \{\mathcal{G}_\rho | \dim(P_\rho)_x > d\} ,$$

where $\mathcal{G}_\rho \subset \mathcal{F} | U$ are coherent analytic subsheaves such that $0_x = \cap_{\rho-1}^r (\mathcal{G}_\rho)_x$ is a primary decomposition of the zero-submodule 0_x of $\mathcal{F}_x$ and $(P_\rho)_x$ are the subvariety germs associated to $\mathcal{F}$ at x.

Proof. If $\dim (P_\rho)_x > d$, then for any subvariety germ A of dimension $\leqq d$ at x we have $\mathcal{H}_A^0(\mathcal{F}/\mathcal{G}_\rho)_x = 0$ as in the proof of Lemma (2.1). Hence $\mathcal{H}_d^0(\mathcal{F}/\mathcal{G}_\rho) = 0$ on some open neighborhood of x. From the exact sequence

$$0 \to \mathcal{H}_d^0 \mathcal{G}_\rho \to \mathcal{H}_d^0 \mathcal{F} \to \mathcal{H}_d^0(\mathcal{F}/\mathcal{G}_\rho)$$

we obtain

$$\mathcal{H}_d^0 \mathcal{F} = \mathcal{H}_d^0 \mathcal{G}_\rho \subset \mathcal{G}_\rho$$

on some open neighborhood of x and so

$$\mathcal{H}_d^0 \mathcal{F} \subset \cap \{\mathcal{G}_\rho \mid \dim(P_\rho)_x > d\} .$$

Conversely, if $s_x \in \mathcal{F}_x$ is contained in this intersection, then for any $(P_\sigma)_x$ with dim $(P_\sigma)_x \leqq d$ we have

$$\prod_\sigma (\mathfrak{P}_\sigma)_x^{k_\sigma} s_x = 0$$

for some k_σ as in Lemma (2.1), where $\mathfrak{P}_\sigma$ is the ideal-sheaf of the subvariety germ $(P_\sigma)_x$. Hence dim Supp $s_x \leqq d$ and $s_x \in (\mathcal{H}_d^0 \mathcal{F})_x$.

Q.E.D.

(2.6) Theorem. Let $\mathcal{F}$ be a coherent analytic sheaf on a complex space X and let $P_{\rho x}(\mathcal{F})$ $(1 \leqq \rho \leqq r)$ be the subvariety germs associated to $\mathcal{F}$ at x. Then the germ at x of the union of d-dimensional branches of $S_d(\mathcal{F})$ is equal to the union of all d-dimensional $P_{\rho x}(\mathcal{F})$.

Proof. Let $I_d \subset \{1,\ldots,r\}$ be the set of all ρ such that dim $P_{\rho x}(\mathcal{F}) = d$. By Corollary (2.4) we need only prove that

$$\mathrm{Supp}(\mathcal{H}_d^0 \mathcal{F}/\mathcal{H}_{d-1}^0 \mathcal{F})_x = \cup \{P_{\rho x}(\mathcal{F}) \mid \rho \in I_d\}.$$

Let $\mathcal{A}$ be the ideal-sheaf of $S := \mathrm{Supp}(\mathcal{H}_d^0 \mathcal{F}/\mathcal{H}_{d-1}^0 \mathcal{F})$. By Lemma (2.5) we can find an open neighborhood U of x such that on U

$$\mathcal{H}_{d-1}^0 \mathcal{F} = \mathcal{H}_d^0 \mathcal{F} \cap (\cap \{\mathcal{Q}_\rho \mid \rho \in I_d\}),$$

where $\mathcal{Q}_\rho \subset \mathcal{F} \mid U$ are coherent analytic subsheaves such that $0_x = \cap_{\rho=1}^r (\mathcal{Q}_\rho)_x$ is a primary decomposition of the zero-submodule 0_x of $\mathcal{F}_x$. If $\rho \in I_d$, then any $f_x \in \mathcal{A}_x$ is a zero-divisor of $(\mathcal{F}/\mathcal{Q}_\rho)_x$, because $f_x^k(\mathcal{H}_d^0 \mathcal{F})_x \subset (\mathcal{Q}_\rho)_x$ for some k and because $(\mathcal{H}_d^0 \mathcal{F})_x \not\subset (\mathcal{Q}_\rho)_x$. Hence

$$\mathcal{A}_x \subset \cap \{(\mathcal{P}_\rho)_x | \rho \in I_d\} ,$$

where $\mathcal{P}_\rho$ is the ideal-sheaf of the subvariety germ $P_{\rho x}(\mathcal{F})$. Consequently

$$S_x \supset \cup \{P_{\rho x}(\mathcal{F}) \mid \rho \in I_d\} .$$

On the other hand,

$$\Pi\{(\mathcal{P}_\rho)_x^{k_\rho} \mathcal{F}_x | \rho \in I_d\} \subset \cap \{(\mathcal{Q}_\rho)_x | \rho \in I_d\}$$

for some $k_\rho (\rho \in I_d)$. Hence

$$\Pi\{(\mathcal{P}_\rho)_x^{k_\rho}(\mathcal{H}_d^0 \mathcal{F}/\mathcal{H}_{d-1}^0 \mathcal{F})_x | \rho \in I_d\} = 0$$

and we obtain

$$S_x \subset \cup \{P_{\rho x}(\mathcal{F}) | \rho \in I_d\} .$$

Q.E.D.

<u>Remark</u>. If $\mathcal{F}$ has no $P_{\rho x}(\mathcal{F})$ of dimension $\leqq d$, then $S_d(\mathcal{F})$ need not be empty, because Theorem (2.6) only states that $S_k(\mathcal{F})$ has no k-

dimensional branches for $k \leqq d$.

Let $\mathfrak{F} \subset \mathfrak{G}$ be coherent analytic sheaves on a complex space X. For an integer $d \geqq 0$ we define the subsheaf $\mathfrak{F}_d \subset \mathfrak{G}$ by the presheaf

$$U \to \varinjlim\{\Gamma(U, \mathfrak{F}[A]) \mid A \in \mathfrak{A}_d(U)\}$$

$$= \{s \in \Gamma(U, \mathfrak{G}) \,\big|\, s|U-A \in \Gamma(U-A, \mathfrak{F}) \text{ for some } A \in \mathfrak{A}_d(U)\}.$$

$\mathfrak{F}_d$ is called the dth <u>relative gap-sheaf of $\mathfrak{F}$ in $\mathfrak{G}$</u>. We have the ascending chain

$$\mathfrak{F} \subset \mathfrak{F}_0 \subset \mathfrak{F}_1 \subset \ldots \subset \mathfrak{G}.$$

If we consider again the canonical sequence

$$0 \to \mathfrak{F} \to \mathfrak{G} \xrightarrow{\pi} \mathfrak{G}/\mathfrak{F} \to 0,$$

we have

$$\mathfrak{F}_d = \pi^{-1}\mathcal{H}^0_d(\mathfrak{G}/\mathfrak{F}) = \pi^{-1}\mathcal{H}^0_{S_d(\mathfrak{G}/\mathfrak{F})}(\mathfrak{G}/\mathfrak{F}) = \mathfrak{F}[S_d(\mathfrak{G}/\mathfrak{F})]$$

From Lemma (2.5) we obtain the following theorem.

(2.7) <u>Theorem</u>. If $\mathfrak{F} \subset \mathfrak{G}$ are coherent analytic sheaves on a complex space, then for any integer $d \geqq 0$ the sheaf $\mathfrak{F}_d$ is coherent and is equal to $\mathfrak{F}[S_d(\mathfrak{G}/\mathfrak{F})]$. Moreover, if $\mathfrak{F}_x = \bigcap_{\rho=1}^r Q_\rho$ is a primary decomposition of $\mathfrak{F}_x$ in $\mathfrak{G}_x$ and $P_{\rho x} = P_{\rho x}(\mathfrak{G}/\mathfrak{F})$, then there exist an open neighborhood U of x and coherent analytic subsheaves

$\mathcal{O}_\rho \subset \mathcal{G}|U$ such that $(\mathcal{O}_\rho)_x = Q_\rho$ and

$$\mathcal{F}_d|U = \cap \{\mathcal{O}_\rho | \dim P_{\rho x} > d\} \ .$$

(2.8) **Corollary.** For every $x \in X$ the germ of $\mathrm{Supp}(\mathcal{F}_d/\mathcal{F}_{d-1})$ at x is the union of all d-dimensional subvariety germs $P_{\rho x}(\mathcal{G}/\mathcal{F})$ associated to $\mathcal{G}/\mathcal{F}$ at x.

Proof. Follows from Theorem (2.6), Corollary (2.4) and the isomorphism

$$\mathcal{F}_d/\mathcal{F}_{d-1} \approx \mathcal{H}_d^0(\mathcal{G}/\mathcal{F})/\mathcal{H}_{d-1}^0(\mathcal{G}/\mathcal{F}).$$

Q.E.D.

(2.9) **Corollary.** The following four conditions are equivalent.

(a) $(\mathcal{F}|U)[A] = \mathcal{F}|U$ for any open subset U of X and any $A \in \mathcal{O}_d(U)$.

(b) $\mathcal{F}_d = \mathcal{F}$.

(c) $\mathcal{H}_d^0(\mathcal{G}/\mathcal{F}) = 0$.

(d) $\mathcal{G}/\mathcal{F}$ has no associated prime ideals of dimension $\leqq$ d.

We are going to prove some identity theorems for sections in coherent analytic sheaves. $_n\mathcal{O}$ denotes the sheaf of germs of holomorphic functions on $\mathbb{C}^n$.

.

(2.10) <u>Lemma</u>. Let $z_1,\ldots,z_n \in {}_n\mathcal{O}_0$ be the germs of the coordinate functions and $1 \leqq q < n$. Let $\mathfrak{a} = \Sigma_{i=1}^{q}\, {}_n\mathcal{O}_0\, z_i$, $\mathfrak{b} = \Sigma_{i=q+1}^{n}\, {}_n\mathcal{O}_0 z_i$, and $\mathfrak{m}$ be the maximal ideal of ${}_n\mathcal{O}_0$. Then, for any natural number k,

(a) $\mathfrak{a} \cap \mathfrak{b}^k = \mathfrak{a}\,\mathfrak{b}^k$ and

(b) $\mathfrak{a} \cap \mathfrak{m}^k = \mathfrak{a}\,\mathfrak{m}^{k-1}$.

Proof. (a) Let $f \in \mathfrak{a} \cap \mathfrak{b}^k$ and let

$$f = \Sigma\, a_{\nu_1\ldots\nu_n}\, a_1^{\nu_1}\ldots z_n^{\nu_n}$$

be its power series expansion. Since $f \in \mathfrak{a}$, $a_{\nu_1\ldots\nu_n} = 0$ if $\nu_1 + \ldots + \nu_q = 0$. Since $f \in \mathfrak{b}^k$, $a_{\nu_1\ldots\nu_n} = 0$ if $\nu_{q+1} + \ldots + \nu_n < k$.

Hence

$$f = \sum_{\nu_1+\ldots+\nu_q \geqq 1} \Big(\sum_{\nu_{q+1}+\ldots+\nu_n \geqq k} a_{\nu_1\ldots\nu_n} z_{q+1}^{\nu_{q+1}} \ldots z_n^{\nu_n} \Big) z_{z_1}^{\nu_1} \ldots z_{z_q}^{\nu_q}$$

belongs to $\mathfrak{a}\,\mathfrak{b}^k$.

(b) We may assume $k \geqq 1$. Let

$$\Sigma_{\lambda=1}^{q} f_\lambda z_\lambda = \sum_{\nu_1+\ldots+\nu_n = k} a_{\nu_1\ldots\nu_n}\, z_1^{\nu_1}\ldots z_n^{\nu_n} \in \mathfrak{a} \cap \mathfrak{m}^k ,$$

where $a_{\nu_1\ldots\nu_n} \in {}_n\mathcal{O}_0$. For $1 \leqq \lambda \leqq q$ let

$$g_\lambda = \Sigma \{a_{\nu_1 \ldots \nu_n} \frac{z_1^{\nu_1} \ldots z_n^{\nu_n}}{z_\lambda} \Big| \nu_1 = \ldots = \nu_{\lambda-1} = 0,\ \nu_\lambda \geqq 1\} .$$

Then

$$\Sigma_{\lambda=1}^{q} z_\lambda (f_\lambda - g_\lambda) = \sum_{\nu_{q+1}+\ldots+\nu_n = k} a_{0\ldots 0\, \nu_{q+1}\ldots\nu_n} z_{q+1}^{\nu_{q+1}} \ldots z_n^{\nu_n} .$$

By (a)

$$\Sigma_{\lambda=1}^{q} z_\lambda (f_\lambda - g_\lambda) \in \mathfrak{a}\, \mathfrak{b}^k$$

and we have

$$\Sigma_{\lambda=1}^{q} z_\lambda (f_\lambda - g_\lambda) = \Sigma_{\lambda=1}^{q} z_\lambda h_\lambda ,$$

where $h_\lambda \in \mathfrak{b}^k \subset \mathfrak{m}^k$. Hence

$$\Sigma_{\lambda=1}^{q} z_\lambda f_\lambda = \Sigma_{\lambda=1}^{q} z_\lambda (g_\lambda + h_\lambda) \in \mathfrak{a}\, \mathfrak{m}^{k-1} .$$

Q.E.D.

(2.11) Lemma. Let X be a domain in $\mathbb{C}^n$, $A = X \cap \{z_1 = \ldots = z_q = 0\}$, and $\mathcal{A}$ be the maximal ideal-sheaf of A. Let $\mathfrak{F}$ be a coherent analytic sheaf on X such that Supp $\mathfrak{F} \subset A$ and $\mathcal{A}^k \mathfrak{F} = 0$. Let $s \in \Gamma(X, \mathfrak{F})$. Then dim Supp $s < n - q$ if one of the following two conditions holds.

(a) $s_x \in \mathfrak{m}_x^k \mathfrak{F}_x$ for every $x \in A$, where $\mathfrak{m}_x$ is the maximal ideal of ${}_n\mathcal{O}_x$.

(b) For every $x \in A$, $s_x \in \mathcal{b}_x \mathcal{F}_x$, where $\mathcal{b}_x$ is the ideal generated by $(z_{q+1} - z_{q+1}(x))_x, \dots, (z_n - z_n(x))_x$ in ${}_n\mathcal{O}_x$.

Proof. Let $\mathcal{F}^{(\nu)} = \mathcal{A}^\nu \mathcal{F} / \mathcal{A}^{\nu+1} \mathcal{F}$ and let $B = \bigcup_{\nu=1}^{k} S_{n-q-1} \mathcal{F}^{(\nu)}$. If $x \notin B$, then there exists an open neighborhood U of x and sections

$$s_1^{(\nu)}, \dots, s_{p_\nu}^{(\nu)} \in \Gamma(U, \mathcal{A}^\nu \mathcal{F})$$

generating $\mathcal{A}^\nu \mathcal{F} | U$ such that the induced homomorphism $({}_n\mathcal{O}/\mathcal{A})^{p_\nu} \to \mathcal{F}^{(\nu)}$ is an isomorphism.

(a) We are going to prove by induction on ν $(0 \leqq \nu \leqq k)$ that $s_y \in \mathfrak{m}_y^{k-\nu} (\mathcal{A}^\nu \mathcal{F})_y$ for $y \in U \cap A$. When $\nu = k$, the statement is reduced to $s | U \cap A = 0$ and it follows that Supp $s \subset B$. Suppose, for a fixed $0 \leqq \nu < k$, $s_y \in \mathfrak{m}_y^{k-\nu} (\mathcal{A}^\nu \mathcal{F})_y$ for $y \in U \cap A$. Let $\bar{s}$ denote the image of s in $\Gamma(U, \mathcal{F}^{(\nu)})$. Then $\bar{s}_y \in \mathfrak{m}_y^{k-\nu} \mathcal{F}_y^{(\nu)}$. Since as an ${}_n\mathcal{O}/\mathcal{A}$ -module $\mathcal{F}^{(\nu)}$ is free on $U \cap A$, it follows that $\bar{s} | U \cap A = 0$. If we write

$$s_y = \Sigma_{i=1}^{p_\nu} f_{iy} (s_i^{(\nu)})_y,$$

where $f_{iy} \in \mathfrak{m}_y^{k-\nu}$, then $f_{iy} \in \mathfrak{m}_y^{k-\nu} \cap \mathcal{A}_y$ and, by Lemma (2.10)(b), $f_{iy} \in \mathfrak{m}_y^{k-\nu-1} \mathcal{A}_y$. Hence $s_y \in \mathfrak{m}_y^{k-(\nu+1)} \mathcal{A}_y^{\nu+1} \mathcal{F}_y$ and the induction is complete.

(b) Since $\mathfrak{b}_y({}_n\mathcal{O}/\mathcal{A})_y = \mathfrak{m}_y({}_n\mathcal{O}/\mathcal{A})_y$, by using (2.10) (a) instead of (2.10)(b), in precisely the same manner as in (a) we can prove by induction on ν $(0 \leqq \nu \leqq k)$ that $s_y \in \mathfrak{b}_y(\mathcal{A}^\nu \mathcal{F})_y$ for $y \in U \cap A$. Hence $s|U \cap A = 0$ and Supp $s \subset B$. Q.E.D.

(2.12) Theorem. Let X be a domain in $\mathbb{C}^n$, $\mathcal{F}$ be a coherent analytic sheaf on X, and $s \in \Gamma(X, \mathcal{F})$. Assume that $s_x \in \mathfrak{a}_x^{(q)} \mathcal{F}_x$ for every $x \in X$ and every ideal $\mathfrak{a}_x^{(q)} \subset {}_n\mathcal{O}_x$ whose subvariety germ at x is a q-dimensional plane parallel to q coordinate axes. Then dim Supp $s < n - q$.

Proof. Take arbitrarily $x_0 \in X$. We can find an open neighborhood U of x_0, coherent analytic subsheaves $\mathcal{Q}_\rho \subset \mathcal{F}|U$, and coherent ideal-sheaves $\mathcal{P}_\rho \subset {}_n\mathcal{O}\ |U$ $(1 \leqq \rho \leqq r)$ such that $0_{x_0} = \bigcap_{\rho=1}^r (\mathcal{Q}_\rho)_{x_0}$ is a primary decomposition of the zero-submodule 0_{x_0} of $\mathcal{F}_{x_0}$ and $(\mathcal{P}_\rho)_{x_0}$ $(1 \leqq \rho \leqq r)$ are the associated prime ideals of 0_{x_0}. By shrinking U, we can assume that $\mathcal{P}_\rho$ is the ideal-sheaf of the subvariety P_ρ defined by $\mathcal{P}_\rho$ and that dim P_ρ is the same as the dimension d_ρ of $(P_\rho)_{x_0}$. By Lemma (2.5) it suffices to prove that $s_{x_0} \in (\mathcal{Q}_\rho)_{x_0}$ for $d_\rho \geqq n - q$.

By using induction on q, we conclude that we need only prove the case $d_\rho = n - q$. $(P_\rho, {}_n\mathcal{O}/\mathcal{P}_\rho)$ is a reduced complex space. Take an arbitrary regular point x of P_ρ. By permuting the coordinates of $\mathbb{C}^n$, we can assume that the projection $P_\rho \to \mathbb{C}^{n-q}$ defined by $z_{q+1}, \dots, z_n$ has rank n-q at x. It follows that

$$\dim_x P_\rho \cap \{z_{q+1} = z_{q+1}(x), \ldots, z_n = z_n(x)\} = 0.$$

By applying a transformation of the local coordinates at x which leaves the coordinates $z_{q+1}, \ldots, z_n$ fixed, we may assume that, for some open neighborhood V of x,

$$P_\rho \cap V = \{z_1 = \ldots = z_q = 0\} \cap V.$$

If $\overline{s} \in \Gamma(P_\rho \cap V, \mathcal{F}/\mathcal{Q}_\rho)$ denotes the image of s, then we have $\overline{s}_y \in \mathcal{b}_y(\mathcal{F}/\mathcal{Q}_\rho)_y$ for $y \in V \cap P_\rho$, where $\mathcal{b}$ is the ideal-sheaf generated by $z_{q+1} - z_{q+1}(x), \ldots, z_n - z_n(x)$. By Lemma (2.11)(b) we have

$$\overline{s} \in \Gamma(P_\rho, \mathcal{H}^0_{n-(q+1)}(\mathcal{F}/\mathcal{Q}_\rho)).$$

Since $(\mathcal{Q}_\rho)_{x_0}$ is a primary submodule of $\mathcal{F}_{x_0}$, we can assume that at the beginning U is chosen so small that $\mathcal{H}^0_{n-(q+1)}(\mathcal{F}/\mathcal{Q}_\rho) = 0$. Hence $\overline{s}|U = 0$ and $s_{x_0} \in (\mathcal{Q}_\rho)_{x_0}$. Q.E.D.

(2.13) Corollary. Suppose $\mathcal{F} \subset \mathcal{G}$ are two coherent analytic sheaves with $\mathcal{F}_{n-(q+1)} = \mathcal{F}$ in $\mathcal{G}$ and suppose $s \in \Gamma(X, \mathcal{G})$. Then $s \in \Gamma(X, \mathcal{F})$ if, for every $x \in X$, $s_x \in \mathcal{F}_x + \mathfrak{a}_x^{(q)} \mathcal{G}_x$, i.e. if the analytic restriction of s to every q-dimensional plane which is parallel to q coordinate axes is in the analytic restriction of $\mathcal{F}$.

Let $(X, \mathcal{O})$ be a general complex space, $x \in X$, and $\mathcal{F}$ be a coherent analytic sheaf on X. If $0_x = \bigcap_{\rho=1}^r (\mathcal{Q}_\rho)_x$ is a primary decomposition of the zero-submodule 0_x of $\mathcal{F}_x$ and $(\mathcal{P}_\rho)_x \subset \mathcal{O}_x (1 \leqq \rho \leqq r)$ are the associated prime ideals of 0_x, then we define the order of $\mathcal{F}$ at x as

$$\operatorname{ord}_x \mathcal{F} = \operatorname{Min}\{\nu \mid (\mathcal{P}_\rho)_x^\nu \, \mathcal{F}_x \subset (\mathcal{Q}_\rho)_x \text{ for } 1 \leqq \rho \leqq r\}.$$

Let $\mathfrak{m}_x$ be the maximal ideal of $\mathcal{O}_x$.

(2.14) Theorem. Let $x_0 \in X$ and $k_0 = \operatorname{ord}_{x_0} \mathcal{F}$. If $s \in \Gamma(X, \mathcal{F})$ and G is an open neighborhood of x_0 such that $s_x \in \mathfrak{m}_x^{k_0} \mathcal{F}_x$ for $x \in G$, then $s_{x_0} = 0$. If Q is a relatively compact open subset of X, then there is an integer depending only on Q such that, if $v \in \Gamma(W, \mathcal{F})$ for some open subset W of Q and $s_x \in \mathfrak{m}_x^k \mathcal{F}_x$ for $x \in W$, then $v = 0$.

Proof. Since the theorem is local in nature, by embedding X locally as a complex subspace of a domain in $\mathbb{C}^n$, we can assume without loss of generality that X is a domain in $\mathbb{C}^n$ and $G = X$. As in the proof of Theorem (2.12) we have U, $\mathcal{Q}_\rho$, $\mathcal{P}_\rho$, P_ρ, and d_ρ. By shrinking U, we may assume that on U we have $0 = \bigcap_{\rho=1}^r \mathcal{Q}_\rho$, $\mathcal{P}_\rho^{k_0} \mathcal{F} \subset \mathcal{Q}_\rho$, and $\mathcal{H}^0_{d_\rho - 1}(\mathcal{F}/\mathcal{Q}_\rho) = 0$.

Suppose $u \in \Gamma(D, \mathcal{F})$ for some open subset D of U and $u_x \in \mathfrak{m}_x^{k_0} \mathcal{F}_x$ for $x \in D$. The theorem follows if we can prove that $u = 0$. Fix an arbitrary ρ. For every regular point x of $P_\rho \cap D$ there exists an open neighborhood V of x in D such that, after a transformation of local coordinates at x,

$$P_\rho \cap V = \{z_{d_\rho+1} = \dots = z_n = 0\} \cap V.$$

Let $\bar{u} \in \Gamma(V, \mathcal{F}/\mathcal{Q}_\rho)$ be the image of u. Then $\bar{u}_y \in \mathfrak{m}_y^{k_0}(\mathcal{F}/\mathcal{Q}_\rho)_y$ for $y \in V$. By Lemma (2.11)(a), $\bar{u} \in \Gamma(V, \mathcal{H}^0_{d_\rho - 1}(\mathcal{F}/\mathcal{Q}_\rho))$. Since x is an arbitrary regular point of $P_\rho \cap D$,

$$\bar{u} \in \Gamma(D, \mathcal{H}^0_{d_\rho - 1}(\mathcal{F}/\mathcal{Q}_\rho)) = 0 \ .$$

$u \in \Gamma(D, \mathcal{Q}_\rho)$. Hence $u = 0$. Q.E.D.

§3 Sheaves of local cohomology and absolute gap-sheaves

The relative gap-sheaves of §2 which can only be formed for subsheaves are insufficient for many purposes. So it is desirable to investigate the coherence of the so-called absolute gap-sheaves, some of which are already introduced in (0.4) and denoted by $\mathcal{R}_A^0 \mathcal{F}$ or $\mathcal{R}_A^1 \mathcal{F}$. Since they are related to the sheaves of local cohomology by the exact sequences

$$0 \to \mathcal{H}_A^0 \mathcal{F} \to \mathcal{F} \to \mathcal{R}_A^0 \mathcal{F} \to \mathcal{H}_A^1 \mathcal{F} \to 0$$

$$0 \to \mathcal{R}_A^i \mathcal{F} \overset{\approx}{\to} \mathcal{H}_A^{i+1} \mathcal{F} \to 0,$$

we investigate the sheaves $\mathcal{H}_A^i \mathcal{F}$ for which we can characterize the coherence property. The result (3.5) is an analogue to a theorem of A. Grothendieck [8, Cor. VIII-II-3].

(3.1) <u>Proposition</u>. Let A be a subvariety in a complex space $(X, \mathcal{O})$ and $\mathcal{F}$ be a coherent analytic sheaf on X. If $\mathcal{H}_A^i \mathcal{F}$ is coherent for $i \leq q$, then

$$\dim A \cap \overline{S}_{k+q+1}(\mathcal{F} | X - A) \leq k$$

for every k, where $\overline{S}_m(\mathcal{F}|X-A)$ denotes the topological closure of the subvariety $S_m(\mathcal{F}|X-A)$ in X (which is equal to the union of all branches of $S_m(\mathcal{F})$ not contained in A).

Proof. If $q = 0$, then

$$\bar{S}_m(\mathcal{F}|X-A) \subset S_m(\mathcal{F}/\mathcal{H}^0_A\mathcal{F})$$

and, since $\mathcal{H}^0_A(\mathcal{F}/\mathcal{H}^0_A\mathcal{F}) = 0$, in this case the proposition follows from Theorem (1.14).

If $q = 1$, then $\mathcal{R}^0_A\mathcal{F}$ is coherent and the result follows from

$$\bar{S}_m(\mathcal{F}|X-A) \subset S_m(\mathcal{R}^0_A\mathcal{F})$$

and

$$\mathcal{H}^i_A\mathcal{R}^0_A\mathcal{F} = 0 \text{ for } i = 0,1 .$$

So, for the proof, by induction we may assume $q \geqq 2$ and that we have already proved

$$\dim A \cap \bar{S}_{k+q}(\mathcal{F}|X-A) \leqq k$$

for every k.

Assume that there is a point $x \in A$ such that

$$\dim_x A \cap \bar{S}_{k+q+1}(\mathcal{F}|X-A) = k + 1 .$$

Then we can find a point

$$y \in A \cap \bar{S}_{k+q+1}(\mathcal{F}|X-A)$$

such that $y \notin \overline{S}_{k+q}(\mathcal{F} | X-A)$ and

$$(*) \qquad \dim_y A \cap \overline{S}_{k+q+1}(\mathcal{F} | X-A) = k + 1.$$

For some open neighborhood U of y,

$$U \cap \overline{S}_{k+q}(\mathcal{F} | X-A) = \emptyset.$$

Since we have (*) and no branch of $\overline{S}_{k+q+1}(\mathcal{F} | X-A)$ is contained in A,

$$\dim_y \overline{S}_{k+q+1}(\mathcal{F} | X-A) \geqq k + 2.$$

Hence, by shrinking U, we can find a (k+2)-dimensional subvariety B in $U \cap \overline{S}_{k+q+1}(\mathcal{F} | X-A)$ such that $\dim A \cap B = k + 1$ and $B = (B-A)^{-}$. Since $q \geqq 2$, by shrinking U again, we can find $f \in \Gamma(U, \mathcal{O})$ such that the subvariety V(f) defined by f contains B but does not contain any $(\ell+q+1)$-dimensional branch of $\overline{S}_{\ell+q+1}(\mathcal{F} | X-A)$ for $\ell \geqq k$. So we have

$$\dim V(f) \cap \overline{S}_{\ell+q+1}(\mathcal{F} | X-A) \leqq \ell + q$$

for $\ell \geqq k$. Since

$$U \cap \overline{S}_{k+q}(\mathcal{F} | X-A) = \emptyset,$$

we obtain

$$\dim V(f) \cap S_{m+1}(\mathcal{F}) \cap (U-A) \leqq m$$

for every m. By Corollary (1.18) f_z is not a zero-divisor of $\mathfrak{F}_z$ for $z \in U-A$. The sheaf-homomorphism $\alpha: \mathfrak{F} \to \mathfrak{F}$ defined by multiplication by f is injective on U - A. The exact sequence

$$0 \to \mathfrak{F}|U - A \xrightarrow{\alpha} \mathfrak{F}|U - A \to (\mathfrak{F}/f\mathfrak{F})|U - A \to 0$$

yields the exact sequence

$$\dots \to \mathcal{R}_A^i \mathfrak{F} \to \mathcal{R}_A^i \mathfrak{F} \to \mathcal{R}_A^i(\mathfrak{F}/f\mathfrak{F}) \to \mathcal{R}_A^{i+1}\mathfrak{F} \to \dots .$$

Since $q \geqq 2$ and $\mathcal{R}_A^i \mathfrak{F}$ are coherent for $i \leqq q - 1$, it follows that $\mathcal{R}_A^i(\mathfrak{F}/f\mathfrak{F})$ is coherent for $i \leqq q - 2$ or $\mathcal{H}_A^i(\mathfrak{F}/f\mathfrak{F})$ is coherent for $i \leqq q - 1$. By induction hypothesis we have

$$\dim A \cap \overline{S}_{\ell+q}((\mathfrak{F}/f\mathfrak{F})|U-A) \leqq \ell$$

for every ℓ. Since

$$B - A \subset S_{k+q+1}(\mathfrak{F}|X-A)$$

and

$$\operatorname{codh}_z(\mathfrak{F}/f\mathfrak{F}) = \operatorname{codh}_z \mathfrak{F} - 1$$

for every $z \in B - A \subset V(f)$, we have

$$B - A \subset S_{k+q}((\mathfrak{F}/f\mathfrak{F})|U-A).$$

Hence we obtain dim $A \cap B \leqq k$, which is a contradiction to the choice of B. Q.E.D.

(3.2) <u>Lemma</u>. If $\mathcal{H}^i_A \mathcal{F}$ is coherent for $i \leqq q$, then for any Stein open subset Ω of X the canonical homomorphism

$$H^i_A(\Omega, \mathcal{F}) \to \Gamma(\Omega, \mathcal{H}^i_A \mathcal{F})$$

is an isomorphism for $i \leqq q + 1$.

Proof. Follows from Lemma (0.6) and Cartan's theorem B. Q.E.D.

We are now going to prove that the conditions of Proposition (3.1) are also sufficient for the coherence of $\mathcal{H}^i_A \mathcal{F}$. For the proof we need two lemmas.

(3.3) <u>Lemma</u>. Let D be a domain in $\mathbb{C}^n$, $A \subset D$ a subvariety of dimension $\leqq d$, and $\mathcal{F}$ a coherent analytic sheaf on D. For $0 \leqq q \leqq n - d$ let $\mathcal{J}_q$ be the ideal-sheaf of $S_{q+d-1}(\mathcal{F})$. Then, for any open subset $X \subset\subset D$ whose topological closure is holomorphically convex, there exists an integer k such that for any open subset U of X

$$\Gamma(U, \mathcal{J}_q)^k \, H^i_A(U, \mathcal{F}) = 0$$

for $i < q$.

Proof. First we show by descending induction on q that we can assume $q = n - d$. Let $q < n - d$. Since X^- is holomorphically convex, there exists an exact sequence

$$0 \to \mathcal{K} \to {}_n\mathcal{O}^s \to \mathcal{F} \to 0$$

on an open neighborhood of X^-.

$$S_{q+d-1}(\mathcal{F}) = S_{(q+1)+d-1}(\mathcal{K})$$

and both have $\mathcal{J}_q$ as ideal-sheaf. If $i < q$, then $\mathcal{H}^j_A({}_n\mathcal{O}) = 0$ for $j \leqq i + 1$. By Lemma (0.6) for any open subset U of X we have $H^i_A(U, {}_n\mathcal{O}) = 0$ $(i \leqq q)$ and so

$$H^i_A(U, \mathcal{F}) = H^{i+1}_A(U, \mathcal{K}) \quad (i < q).$$

Hence, for $i < q$, $H^i_A(U, \mathcal{F})$ is annihilated by $\Gamma(U, \mathcal{J}_q)^k$ if and only if $H^{i+1}_A(U, \mathcal{K})$ is.

Now let $q = n - d$. Since X^- is holomorphically convex, there is an exact sequence

$$\cdots \to {}_n\mathcal{O}^{s_q} \to \cdots \to {}_n\mathcal{O}^{s_0} \to \mathcal{F}^* \to 0$$

on an open neighborhood of X^-, where $\mathcal{F}^* = \mathcal{H}om_{{}_n\mathcal{O}}(\mathcal{F}, {}_n\mathcal{O})$. By applying the functor * to this sequence, we obtain the following sequence of sheaf-homomorphisms:

$$(\alpha) \qquad 0 \xrightarrow{\varphi_{-2}} \mathcal{F} \xrightarrow{\varphi_{-1}} {}_n\mathcal{O}^{s_0} \xrightarrow{\varphi_0} {}_n\mathcal{O}^{s_1} \xrightarrow{\varphi_1} \cdots,$$

where φ_{-1} is the composite map

$$\mathcal{F} \to \mathcal{F}^{**} \to {}_n\mathcal{O}^{s_0} .$$

Since $\mathcal{F}$ is locally free on $D - S_{n-1}(\mathcal{F})$, (α) is exact on $X - S_{n-1}(\mathcal{F})$. We define for $i \geqq -1$

$$\mathcal{B}_i = \operatorname{Im} \varphi_{i-1} , \quad \mathcal{Z}_i = \operatorname{Ker} \varphi_i ,$$

and $\mathcal{L}_i = \mathcal{Z}_i / \mathcal{B}_i$. Then Supp $\mathcal{L}_i \subset S_{n-1}(\mathcal{F})$. Since X^- is compact, by Hilbert Nullstellensatz we can find an integer ℓ such that $\mathcal{J}_q^{\ell} \mathcal{L}_i | X^- = 0$. Hence for any open subset U of X we have

(β) $$\Gamma(U, \mathcal{J}_q)^{\ell} H_A^k(U, \mathcal{L}_i) = 0 \text{ for } k \geqq 0 .$$

Consider the following exact sequences

(γ) $$\begin{cases} 0 \to \mathcal{Z}_{-1} \to \mathcal{F} \to \mathcal{B}_0 \to 0 \\ 0 \to \mathcal{Z}_j \to {}_n\mathcal{O}^{s_j} \to \mathcal{B}_{j+1} \to 0 \quad (j \geqq 0) \end{cases}$$

(δ) $$0 \to \mathcal{B}_j \to \mathcal{Z}_j \to \mathcal{L}_j \to 0 .$$

We are going to prove the following statements for $j \geqq 0$ by descending induction on j.

$(\varepsilon)_j$ $\Big\{$ There is an integer $\ell \geqq 0$ such that for any open subset U of X, $\Gamma(U, \mathcal{J}_q)^{\ell} H_A^i(U, \mathcal{Z}_j) = 0$ for $i < q - j$.

$(\zeta)_j$ $\left\{\begin{array}{l}\text{There is an integer } \ell \geqq 0 \text{ such that for any open subset U} \\ \text{of X, } \Gamma(U, \mathcal{J}_q)^{\ell} H^i_A(U, \mathcal{B}_j) = 0 \text{ for } i < q - j.\end{array}\right.$

By (δ) we obtain the exact sequence

$$H^{i-1}_A(U, \mathcal{L}_j) \to H^i_A(U, \mathcal{B}_j) \to H^i_A(U, \mathcal{Z}_j)$$

and by (β) we have the implication $(\varepsilon)_j \Rightarrow (\zeta)_j$. We now prove that for $j \geqq 1$ $(\zeta)_j \Rightarrow (\varepsilon)_{j-1}$. Since $H^i_A(U, {}_n\mathcal{O}) = 0$ for $i < q$, if $i < q - (j-1) \leqq q$ we get from the second sequence in (γ) the exact sequences

$$0 \to H^{i-1}_A(U, \mathcal{B}_j) \to H^i_A(U, \mathcal{Z}_{j-1}) \to 0$$

and, since $i - 1 < q - j$, $(\varepsilon)_{j-1}$ follows from $(\zeta)_j$. The induction is complete, because $(\varepsilon)_j$ and $(\zeta)_j$ are both vacuous when $j > q$.

From the first sequence in (γ) we obtain the exact sequence

$$\ldots \to H^i_A(U, \mathcal{Z}_{-1}) \to H^i_A(U, \mathcal{F}) \to H^i_A(U, \mathcal{B}_0) \to \ldots .$$

By $(\varepsilon)_0$, $H^i_A(U, \mathcal{B}_0)$ is annihilated by some $\Gamma(U, \mathcal{J}_q)^{\ell}$ for $i < q$ and, since $\mathcal{Z}_{-1} = \mathcal{L}_{-1}$, the final result follows from (β). Q.E.D.

(3.4) Lemma. Let $D \subset \mathbb{C}^n$ be a domain, $A \subset V$ subvarieties in D, and $\mathcal{F}$ a coherent analytic sheaf on D. Let S'_{k+q} be the union of branches of $S_{k+q}(\mathfrak{F})$ which are not contained in V. Assume that $\dim A \cap S'_{k+q} \leqq k$ for every k. Let $\mathcal{J}$ be the ideal-sheaf of V and let X be a relatively compact open subset of D whose topological closure X^- is holomorphically convex. Then there exists an integer $\ell \geqq 0$ such that for any open subset U of X, $\Gamma(U, \mathcal{J})^\ell H^i_A(U, \mathfrak{F}) = 0$ for $i < q$.

Proof. Let $A_k = A \cap S'_{k+q}$ and $A''_{k+1} = A_{k+1} - S'_{k+q}$. If $k + q = n$, we have $A_k = A$ and $\dim A \leqq n - q$. We are going to prove the lemma for A_k by induction on k.

If $k = 0$, we have $\dim A_0 \leqq 0$ and by Lemma (3.3) the result follows from the fact that, in some open neighborhood of A_0, $S_{q-1}(\mathfrak{F}) \subset V$ and $\mathcal{J} \subset \mathcal{J}(S_{q-1}(\mathfrak{F}))$.

Since

$$S_{q+(k+1)-1}(\mathfrak{F}) \cap (X - S'_{q+k}) \subset V$$

and since $(X - S'_{q+k})^- = X^-$, by Lemma (3.3) we have

$$\Gamma(U - S'_{k+q}, \mathcal{J})^\ell H^i_{A_{k+1}}(U - S'_{q+k}, \mathfrak{F}) = 0$$

for $i < q$ and for some integer $\ell \geqq 0$ which is independent of U. By the excision theorem (0.1)

$$H^i_{A_{k+1}}(U - S'_{q+k}, \mathfrak{F}) = H^i_{A''_{k+1}}(U, \mathfrak{F}) .$$

Since

$$\Gamma(U, \mathcal{J}) \subset \Gamma(U - S'_{q+k}, \mathcal{J}),$$

we get

$$\Gamma(U, \mathcal{J})^{\ell} H^i_{A''_{k+1}}(U, \mathcal{F}) = 0$$

for $i < q$. Now by the exact sequence (0.3)

$$\cdots \to H^{i-1}_{A''_{k+1}}(U, \mathcal{F}) \to H^i_{A_k}(U, \mathcal{F}) \to H^i_{A_{k+1}}(U, \mathcal{F}) \to H^i_{A''_{k+1}}(U, \mathcal{F}) \to \cdots$$

the lemma for A_{k+1} follows from the lemma for A_k and the induction is complete. Q.E.D.

(3.5) Theorem. Let X be a complex space, $A \subset X$ a subvariety, and $\mathcal{F}$ a coherent analytic sheaf on X. Then for $q \geqq 0$ the following two conditions are equivalent.

(a) $\dim A \cap \overline{S}_{k+q+1}(\mathcal{F}|X-A) \leq k$ for every k.

(b) $\mathcal{H}^i_A \mathcal{F}$ is coherent for $0 \leqq i \leqq q$.

Proof. By Proposition (3.1) we need only prove (a) => (b). Since both conditions are local in nature, we may assume that X is a relatively compact subdomain of a domain $D \subset \mathbb{C}^n$, X^- is holomorphically convex, and $\mathcal{F}$ is defined on D. By Lemma (3.4) (with A = V and q + 1 replacing q) we obtain an integer $\ell \geqq 0$ such that

$$\Gamma(U, \mathcal{J})^{\ell}\, H^i_A(U, \mathcal{F}) = 0$$

for $i \leqq q$ and any open subset U of X, where $\mathcal{J}$ is the ideal-sheaf of A. Hence $\mathcal{J}^{\ell}\, \mathcal{H}^i_A \mathcal{F} = 0$.

We may assume that $\mathcal{H}^0_A \mathcal{F} = 0$, because $\mathcal{H}^0_A \mathcal{F}$ is always coherent (Proposition (0.10)) and $\mathcal{H}^i_A \mathcal{F} = \mathcal{H}^i_A(\mathcal{F}/\ \mathcal{H}^0_A \mathcal{F})$, $i \geqq 1$. Take $x \in A$. By Proposition (1.9) we can find an open neighborhood U of x and $f \in \Gamma(U, \mathcal{J})$ such that the sequence

$$0 \to \mathcal{F}|U \overset{\alpha}{\to} \mathcal{F}|U \to (\mathcal{F}/f^{\ell}\mathcal{F})|U \to 0$$

is exact, where α is defined by multiplication by f^{ℓ}. By $\mathcal{J}^{\ell}\, \mathcal{H}^i_A \mathcal{F} = 0$ we obtain the exact sequences

$$0 \to \mathcal{H}^{i-1}_A \mathcal{F} \to \mathcal{H}^{i-1}_A(\mathcal{F}/f^{\ell}\mathcal{F}) \to \mathcal{H}^i_A \mathcal{F} \to 0$$

for $i \leqq q$. Since

$$S_{k+q}(\mathcal{F}/f^{\ell}\mathcal{F}) \subset S_{q+k+1}(\mathcal{F}),$$

we have

$$\dim A \cap \overline{S}_{(q-1)+k+1}(\mathcal{F}/f\mathcal{F}|U-A) \leqq k$$

for every k and hence, by induction on q, we have the coherence of $\mathcal{H}^{i-1}_A(\mathcal{F}/f^{\ell}\mathcal{F})$ and $\mathcal{H}^{i-1}_A \mathcal{F}$ on U for $i \leqq q$. The coherence of $\mathcal{H}^q_A \mathcal{F}$ at x follows. Q.E.D.

(3.6) Corollary. If for every $x \in A$ there is an open neighborhood U of x such that $\mathrm{codh}(\mathcal{F}|U-A) \geq q + \dim_x A + 2$, then $\mathcal{R}^i_A \mathcal{F}$ is coherent for $0 \leq i \leq q$.

(3.7) Remark. The connection between Theorem (3.5) and the corresponding theorem [8, Cor. VIII-II-3] in algebraic geometry can easily be established by the following arguments.

If $(X, \mathcal{O})$ is a complex space, then for every $x \in X$ we denote by $\mathrm{Spec}\ \mathcal{O}_x$ the set of all prime ideals in $\mathcal{O}_x$ equipped with the Zariski topology and denote by $\tilde{\mathcal{O}}_x$ the sheaf of rings on $\mathrm{Spec}\ \mathcal{O}_x$ whose stalk at a prime ideal is the localization of $\mathcal{O}_x$ with respect to that prime ideal. If $A \subset X$ is an analytic subvariety with ideal-sheaf $\mathcal{A}$, then by $V(\mathcal{A}_x)$ we denote the set of all prime ideals containing $\mathcal{A}_x$. $V(\mathcal{A}_x)$ is closed in the Zariski topology of $\mathrm{Spec}\ \mathcal{O}_x$. If $\mathcal{F}$ is a coherent analytic sheaf on X, then we denote by $\tilde{\mathcal{F}}_x$ the coherent $\tilde{\mathcal{O}}_x$-sheaf on $\mathrm{Spec}\ \mathcal{O}_x$ whose stalk at a prime ideal of $\mathcal{O}_x$ is the localization of $\mathcal{F}_x$ with respect to that prime ideal. It is proved in [8] and [10] that

$$(\alpha) \qquad H^i_{V(\mathcal{A}_x)}(\mathrm{Spec}\ \mathcal{O}_x, \tilde{\mathcal{F}}_x) = \varinjlim_n \mathrm{Ext}^i_{\mathcal{O}_x}(\mathcal{O}_x/\mathcal{A}^n_x \mathcal{O}_x, \mathcal{F}_x).$$

Since for any coherent analytic sheaf $\mathcal{F}$ on X we have the identity

$$\mathcal{H}om_{\mathcal{O}}(\mathcal{O}/\mathcal{A}^n\mathcal{O}, \mathcal{F}) = \mathcal{H}om_{\mathcal{O}}(\mathcal{O}/\mathcal{A}^n\mathcal{O}, \mathcal{H}^0_A \mathcal{F}),$$

there is a spectral sequence

$$\varinjlim_n \mathcal{E}xt^p_{\mathcal{O}}(\mathcal{O}/\mathcal{A}^n\mathcal{O},\ \mathcal{H}^q_A\mathcal{F}) \underset{p}{\Rightarrow} \varinjlim_n \mathcal{E}xt^{p+q}_{\mathcal{O}}(\mathcal{O}/\mathcal{A}^n\mathcal{O},\mathcal{F}),$$

which is valid for the stalks as well. If $\mathcal{H}^i_A\mathcal{F}$ are coherent for $i \leqq m$, since the support of $(\mathcal{H}^i_A\mathcal{F})_x$ as a module is contained in $V(\mathcal{A}_x)$ for $i \leqq m$, the spectral sequence degenerates because of (α). So we get the isomorphisms

$$(\mathcal{H}^i_A\mathcal{F})_x = H^i_{V(\mathcal{A}_x)}(\mathrm{Spec}\ \mathcal{O}_x,\ \tilde{\mathcal{F}}_x) \qquad (i \leqq m)$$

and

$$(\beta)\ \varinjlim_n \mathrm{Hom}_{\mathcal{O}_x}(\mathcal{O}_x/\mathcal{A}^n_x\mathcal{O}_x, (\mathcal{H}^{m+1}_A\mathcal{F})_x) = H^{m+1}_{V(\mathcal{A}_x)}(\mathrm{Spec}\ \mathcal{O}_x,\ \tilde{\mathcal{F}}_x).$$

This shows that in the case of the coherence of $\mathcal{H}^i_A\mathcal{F}$ the stalks $(\mathcal{H}^i_A\mathcal{F})_x$ are already determined by $\mathcal{F}_x$. Moreover, from (β) we can derive the following result.

(γ) If for $x \in A$ and $i \leqq q$ the stalks $(\mathcal{H}^i_A\mathcal{F})_x$ are finitely generated over $\mathcal{O}_x$, then there exists a neighborhood U of x such that $\mathcal{H}^i_A\mathcal{F}|U$ are coherent.

The proof follows by induction on q. Since $\mathcal{H}^0_A\mathcal{F}$ is always coherent, the induction basis is given. Now assume that for $m < q$ the sheaves $\mathcal{H}^i_A\mathcal{F}|U$ are already coherent for $i \leqq m$. By (β) the groups

$H^i_{V(\mathcal{A}_x)}(\mathrm{Spec}\ \mathcal{O}_x, \tilde{\mathfrak{F}}_x)$ are finitely generated for $i \leq m+1$, because $(\mathcal{H}^{m+1}_A \mathfrak{F})_x$ is still a finitely generated $\mathcal{O}_x$-module. By [8, Cor. VII-II-3] we can verify canonically that

$$\dim_x A \cap \bar{S}_{(m+1)+k+1}(\mathfrak{F}|U-A) \leq k$$

for every k and this is then valid in some neighborhood of x. By Theorem (3.5), $\mathcal{H}^{m+1}_A \mathfrak{F}|U$ is then coherent too.

One can also establish (γ) in an elementary way. The remark here shows however that the derivation of (γ) from Theorem (3.5) is of purely algebraic nature.

Let X be a complex space and $d \geq 0$ be an integer. As in §2 we consider $\mathfrak{A}_d = \cup \mathfrak{A}_d(U)$, the collection of all locally closed subsets which are analytic and have dimension $\leq d$ in each of their points. For $A \subset B$ in $\mathfrak{A}_d$ there is a canonical homomorphism $\mathcal{H}^i_A \mathfrak{F} \to \mathcal{H}^i_B \mathfrak{F}$, from which we can define the sheaves

$$\mathcal{H}^i_d \mathfrak{F} = \varinjlim \{\mathcal{H}^i_A \mathfrak{F} | A \in \mathfrak{A}_d\} .$$

These sheaves can also be defined by the presheaves

$$U \mapsto \varinjlim \{H^i_A(U, \mathfrak{F}) | A \in \mathfrak{A}_d(U)\} ,$$

where the restriction maps for $U \supset V$ are the limit-homomorphisms of the systems

$$H^i_A(U, \mathcal{F}) \to H^i_A(V, \mathcal{F})$$

for $A \in \mathfrak{A}_d(U)$. In the same way we define the absolute gap-sheaves $\mathcal{R}^i_d \mathcal{F}$ by the presheaves

$$U \mapsto \varinjlim \{H^i(U-A, \mathcal{F}) \mid A \in \mathfrak{A}_d(U)\}.$$

Since $\varinjlim$ is an exact functor, we obtain the exact sequence

$$0 \to \mathcal{H}^0_d \mathcal{F} \to \mathcal{F} \to \mathcal{R}^0_d \mathcal{F} \to \mathcal{H}^1_d \mathcal{F} \to 0$$

and the isomorphisms

$$\mathcal{R}^i_d \mathcal{F} \approx \mathcal{H}^{i+1}_d \mathcal{F}$$

for $i \geqq 1$.

If M is a subset of X, we say that $\dim M \leqq d$, if for every point $x \in M$ there exists an open neighborhood U of x and a subvariety $A \in \mathfrak{A}_d(U)$ such that $M \cap U \subset A$.

(3.8) Theorem. Let X be a complex space, $\mathcal{F}$ a coherent analytic sheaf on X, and $d, q \geqq 0$ two integers. Then the following four conditions are equivalent.

(a) $\dim \operatorname{Supp} \mathcal{H}^i_{d+1} \mathcal{F} \leqq d$ for $i \leqq q$.

(b) $\dim S_{d+q+1}(\mathcal{F}) \leqq d$.

(c) $\mathcal{H}^i_d \mathcal{F}$ is coherent and equal to $\mathcal{H}^i_{S_{d+q+1}(\mathcal{F})} \mathcal{F}$ for $i \leqq q + 1$.

(d) $\mathcal{H}^i_{d+\rho} \mathcal{F}$ is coherent and equal to $\mathcal{H}^i_d \mathcal{F}$ for $i \leqq q + 1 - \rho$.

Proof. (a) => (b). For $x \in \bigcup_{i=1}^q \operatorname{Supp} \mathcal{H}^i_{d+1}\mathcal{F}$ we can find an open neighborhood U of x and an $S \in \mathfrak{A}_d(U)$ such that $\mathcal{H}^i_{d+1}\mathcal{F} | U - S = 0$ for $i \leqq q$. We are going to prove by induction on q that $\mathcal{H}^i_A \mathcal{F} = 0$ for $i \leqq q$ and for any $A \in \mathfrak{A}_{d+1}$ which is contained in U - S. If this is proved, then by Corollary (1.16)

$$\dim (U - S) \cap S_{k+q+1}(\mathcal{F}) \leqq k$$

for $k \leqq d$ and, since $\dim S \leqq d$, (b) follows. If q = 0, it follows directly from the definition of $\mathcal{H}^0_{d+1}(\mathcal{F})$ that $\mathcal{H}^0_A \mathcal{F} = 0$ for any $A \in \mathfrak{A}_{d+1}$ contained in U - S. Assume for $i \leqq p < q$ we have already proved that $\mathcal{H}^i_A \mathcal{F} = 0$ for $A \in \mathfrak{A}_{d+1}$ contained in U - S. Let $y \in U - S$ and V be an open neighborhood of y in U - S. Take $\xi \in H^{p+1}_A(V, \mathcal{F})$ for $A \in \mathfrak{A}_{d+1}(V)$. Since $(\mathcal{H}^{p+1}_{d+1} \mathcal{F})_x = 0$, there exists an open neighborhood W of y in V and $B \in \mathfrak{A}_{d+1}(W)$ such that $A \cap W \subset B$ and ξ is mapped to 0 under the homomorphism $\beta \circ \alpha$ in

$$H^{p+1}_A(V, \mathcal{F}) \xrightarrow{\alpha} H^{p+1}_{A \cap W}(W, \mathcal{F}) \xrightarrow{\beta} H^{p+1}_B(W, \mathcal{F}).$$

Let B' = B - A. Since $\mathcal{H}^i_{B''} \mathcal{F} = 0$ for $i \leqq p$, by Lemma (0.6) we have

$$H^p_{B'}(W, \mathcal{F}) = H^p_B(W - A, \mathcal{F}) = 0 .$$

Hence β is injective and we have $\alpha(\xi) = 0$. This implies that $(\mathcal{H}^{p+1}_A \mathcal{F})_x = 0$. Hence $\mathcal{H}^{p+1}_A \mathcal{F} = 0$ for every $A \in \mathfrak{A}_{d+1}$. The induction is complete and (b) holds.

(b) => (c). Let $S = S_{d+q+1}(\mathcal{F})$. $\dim S \leqq d$. We have $\operatorname{codh}(\mathcal{F}|X - S) \geqq d + q + 2$ and hence by Corollary (3.6) $\mathcal{H}^i_S \mathcal{F}$ is coherent for $i \leqq q + 1$. If $x \in S$ and U is an open neighborhood of x, then for $A \in \mathfrak{A}_d(U)$ we have $\mathcal{H}^i_S \mathcal{F} = \mathcal{H}^i_{A \cup S} \mathcal{F}$, because $\mathcal{H}^i_A \mathcal{F} \, |U - S = 0$. Hence

$$\mathcal{H}^i_S \mathcal{F} = \varinjlim \{ \mathcal{H}^i_A \mathcal{F} \,|\, A \in \mathfrak{A}_d \} = \mathcal{H}^i_d \mathcal{F}$$

and (c) is proved.

(c) => (b). Let $\mathcal{H}^i_d \mathcal{F}$ be coherent for $i \leqq q + 1$ and $S_i = \operatorname{Supp} \mathcal{H}^i_d \mathcal{F}$. We are going to prove that $\dim S_i \leqq d$. Let $x \in S_i$ and let U be an open neighborhood of x such that there exist $s_1, \ldots, s_k \in \Gamma(U, \mathcal{H}^i_d \mathcal{F})$ generating $\mathcal{H}^i_d \mathcal{F} \,|U$. We can find an open neighborhood V of x in U and an $A \in \mathfrak{A}_d(V)$ such that $s_\ell | V$ is induced by some $\xi_\ell \in H^i_A(V, \mathcal{F})$ $(1 \leqq \ell \leqq k)$. It follows that $\operatorname{Supp} \mathcal{H}^i_d \mathcal{F} | V \subset A$ and hence $\dim S_i \leqq d$. Let $S = \bigcup_{i=1}^{q+1} S_i$. Then $\dim S \leqq d$ and $\mathcal{H}^i_d \mathcal{F} | X - S = 0$ for $i \leqq q + 1$. As in the proof of (a) => (b) we obtain $\dim S_{k+(q+1)+1}(\mathcal{F}|X-S) \leqq k$ for $k \leqq d - 1$. In particular, for

$k = d - 1$, we obtain

$\dim S_{d+q+1}(\mathcal{F} \mid X - S) \leqq d - 1 \leqq d$. Since $\dim S \leqq d$, we have

$\dim S_{d+q+1}(\mathcal{F}) \leqq d$.

(b) => (d) follows by writing $S_{d+q+1}(\mathcal{F}) = S_{(d+\rho)+(q-\rho)+1}(\mathcal{F})$ and from the proof of (b) => (c).

(d) => (a) is trivial. Q.E.D.

(3.9) Corollary. The following four conditions are equivalent.

(a) $\dim \operatorname{Supp} \mathcal{H}^0_{d+1}\mathcal{F} \leqq d$.

(b) $\mathcal{H}^0_{d+1}\mathcal{F} = \mathcal{H}^0_d\mathcal{F}$.

(c) $\dim S_{d+1}(\mathcal{F}) \leqq d$.

(d) $\mathcal{R}^0_d\mathcal{F}$ is coherent.

Proof. $\mathcal{R}^0_d\mathcal{F}$ is coherent if and only if $\mathcal{H}^1_d\mathcal{F}$ is coherent.

Q.E.D.

The conclusion (a) => (d) of Corollary (3.9) enables us to construct for a given coherent analytic sheaf $\mathcal{F}$ a coherent analytic sheaf $\mathcal{G}$ such that $\mathcal{F}/\mathcal{H}^0_{d+1}\mathcal{F} \subset \mathcal{G}$ and $\mathcal{R}^0_A\mathcal{G} = \mathcal{G}$ for any $A \in \mathfrak{A}_d$.

Remark. If $\mathcal{H}^i_d\mathcal{F}$ is coherent, then $\mathcal{H}^i_A\mathcal{F}$ need not be coherent for every $A \in \mathfrak{A}_d$, as can be easily seen by comparing Theorems (3.5) and (3.8).

§4 Closedness of coboundary modules

To have a complete description of the groups $H^i_A(X, \mathcal{F})$ in the case where X is Stein and the sheaves $\mathcal{H}^i_A\mathcal{F}$ are coherent, it is desirable to know that the canonical isomorphisms $H^i_A(X, \mathcal{F}) \to \Gamma(X, \mathcal{H}^i_A\mathcal{F})$ are also topological. We are going to show that this is true and indeed it follows from the fact that up to a certain dimension the spaces $H^i_A(X, \mathcal{F})$ are Hausdorff.

(4.1) Lemma. Let Δ be the unit polydisc in $\mathbb{C}^n$ and let $A = \Delta \cap \{z_{d+1} = \dots = z_n = 0\}$, where $0 \leqq d \leqq n - 1$. Then the space $H^{n-d}_A(\Delta, {}_n\mathcal{O})$ is Hausdorff. (Note that $H^i_A(\Delta, {}_n\mathcal{O}) = 0$ for $i \neq n - d$ by Frenkel's lemma (0.14)(F_2).)

Proof. Let $U_i = \Delta \cap \{z_i \neq 0\}$ for $d + 1 \leqq i \leqq n$. Then

$$\mathfrak{U} = \{U_{d+1}, \dots, U_n\}$$

is a Stein covering of $\Delta - A$. Since

$$H^{n-d}_A(\Delta, {}_n\mathcal{O}) \cong H^{n-d-1}(\Delta - A, {}_n\mathcal{O})$$

we need only prove that $H^{n-d-1}(\mathfrak{U}, {}_n\mathcal{O})$ is Hausdorff. Let $U^* = \bigcap_{i=d+1}^n U_i$. Since

$$Z^{n-d-1}(\mathfrak{U}, {}_n\mathcal{O}) = \Gamma(U^*, {}_n\mathcal{O}) ,$$

every $\xi \in Z^{n-d-1}(\mathfrak{U}, {}_n\mathcal{O})$ is a Laurent series

$$\xi = \Sigma^{\infty}_{\nu_{d+1},\ldots,\nu_n = -\infty} a_{\nu_{d+1}\ldots\nu_n}(z_1,\ldots,z_d) z_{d+1}^{\nu_{d+1}} \ldots z_n^{\nu_n} ,$$

where the coefficients $a_{\nu_{d+1}\ldots\nu_n}$ are holomorphic on the unit polydisc of $\mathbb{C}^d$. It is easy to see that $\xi \in B^{n-d-1}(\mathfrak{U}, {}_n\mathcal{O})$ if and only if $a_{\nu_{d+1}\ldots\nu_n} = 0$ for $\nu_{d+1} \leqq -1,\ldots,\nu_n \leqq -1$. Hence by Cauchy's integral formula $B^{n-d-1}(\mathfrak{U}, {}_n\mathcal{O})$ is closed in $\Gamma(U^*, {}_n\mathcal{O})$ equipped with the topology of uniform convergence on compact subsets. $H^{n-d-1}(\mathfrak{U}, {}_n\mathcal{O})$ is a Fréchet space and hence Hausdorff. Q.E.D.

If X is a complex space and $A \subset X$ is a subvariety, then for any coherent analytic sheaf $\mathfrak{F}$ on X we have the exact sequence

$$\ldots \to H^{i-1}(X-A, \mathfrak{F}) \xrightarrow{\delta^{i-1}} H^i_A(X, \mathfrak{F}) \xrightarrow{\alpha} H^i(X, \mathfrak{F}) \to \ldots .$$

We give $H^{i-1}(X-A, \mathfrak{F})$ and $H^i(X, \mathfrak{F})$ the topologies introduced in (0.13). For $i \geqq 1$, we equip $H^i_A(X, \mathfrak{F})$ with the finest topology such that δ^{i-1} is continuous. Then α is continuous, because $\alpha\, \delta^{i-1} = 0$. We equip $H^0_A(X, \mathfrak{F}) \subset \Gamma(X, \mathfrak{F})$ with the topology induced by $\Gamma(X, \mathfrak{F})$. If X is Stein, then $H^i_A(X, \mathfrak{F})$ has the same topology as $H^{i-1}(X-A, \mathfrak{F})$ for $i \geqq 2$ and $H^1_A(X, \mathfrak{F})$ has the quotient topology of $\Gamma(X, \mathfrak{F})/\Gamma(X-A, \mathfrak{F})$.

For an open subset U of X let $N^i_A(U, \mathfrak{F})$ be the topological closure of 0 in $H^i_A(U, \mathcal{F})$. Since the restriction map

$$\beta: H^i_A(U, \mathcal{F}) \to H^i_A(V, \mathcal{F})$$

is continuous for any open subset V of U,

$$\beta(N^i_A(U, \mathcal{F})) \subset N^i_A(V, \mathcal{F}).$$

Define the sheaves $\mathcal{N}^i_A \mathcal{F}$ by the presheaves

$$U \mapsto N^i_A(U, \mathcal{F})$$

with the restriction maps

$$N^i_A(U, \mathcal{F}) \to N^i_A(V, \mathcal{F})$$

induced by β.

(4.2) Lemma. Let D be a domain in $\mathbb{C}^n$ and $A \subset D$ a subvariety of dimension d. Then $\mathcal{N}^i_A({}_n\mathcal{O}) = 0$ for $i \leqq n - d$ and $H^{n-d}_A(D, {}_n\mathcal{O}) = \Gamma(D, \mathcal{H}^{n-d}_A({}_n\mathcal{O}))$ is a Hausdorff space and, when D is Stein, is a Fréchet space.

Proof. Since by Proposition (1.12) $\mathcal{H}^i_A({}_n\mathcal{O}) = 0$ for $i < n - d$, by Lemma (0.6)

$$H_A^{n-d}(U, {}_n\mathcal{O}) = \Gamma(U, \mathcal{H}_A^{n-d}({}_n\mathcal{O}))$$

for every open subset U of D. For $i < n - d$, $\mathcal{N}_A^i({}_n\mathcal{O}) = 0$ follows from $\mathcal{H}_A^i({}_n\mathcal{O}) = 0$. If we can show that $\mathcal{N}_A^{n-d}({}_n\mathcal{O}) = 0$, then $H_A^{n-d}(D, {}_n\mathcal{O})$ is a Hausdorff space, because the commutative diagram

$$\begin{array}{ccc} N_A^{n-d}(D, {}_n\mathcal{O}) & \to & \Gamma(D, \mathcal{N}_A^{n-d}({}_n\mathcal{O})) \\ \downarrow & & \downarrow \\ H_A^{n-d}(D, {}_n\mathcal{O}) & = & \Gamma(D, \mathcal{H}_A^{n-d}({}_n\mathcal{O})) \end{array}$$

implies that $N_A^{n-d}(D, {}_n\mathcal{O}) = 0$.

Let A' be the set of all singular points of A and let A" = A - A'. For $x \in A''$ we can find an open neighborhood U of x in D - A' such that $A \cap U$ is isomorphic to

$$U \cap \{z_{d+1} = \dots = z_n = 0\} ,$$

where Δ is the unit polydisc in $\mathbb{C}^n$. By Lemma (4.1) $H_A^{n-d}(U, {}_n\mathcal{O})$ is Hausdorff. Hence $\mathcal{N}_A^{n-d}({}_n\mathcal{O})$ has support in A' and, by replacing D by D - A', we conclude that $H_A^{n-d}(D - A', {}_n\mathcal{O})$ is Hausdorff. Since $H_{A'}^{n-d}(D, {}_n\mathcal{O}) = 0$, we have the exact sequence

$$0 \to H_A^{n-d}(D, {}_n\mathcal{O}) \xrightarrow{\alpha} H_A^{n-d}(D-A', {}_n\mathcal{O}) .$$

The vanishing of $N_A^{n-d}(D, {}_n\mathcal{O})$ follows from the continuity of α.

When D is Stein, for $d < n$ the Hausdorff space $H_A^{n-d}(D, {}_n\mathcal{O})$ is the continuous open image of $H^{n-d-1}(D - A, {}_n\mathcal{O})$ and therefore $H^{n-d-1}(D - A, {}_n\mathcal{O})$ is Hausdorff and both are Fréchet spaces. When $d = n$, $H_A^{n-d}(D, {}_n\mathcal{O})$ is trivially a Fréchet space. Q.E.D.

(4.3) <u>Lemma</u>. Let D be a domain in $\mathbb{C}^n$, $A \subset D$ a subvariety of dimension $\leqq d$, and $\mathcal{F}$ a coherent analytic sheaf on D. For $0 \leqq q \leqq n - d$ let $\mathcal{J}_q$ be the ideal-sheaf of $S_{q+d-1}(\mathcal{F})$. Then for any open set $X \subset\subset D$ whose topological closure X^- is holomorphically convex there exists an integer $\ell \geqq 0$ such that, for any open subset U of X, $\Gamma(U, \mathcal{J}_q)^{\ell} N_A^q(U, \mathcal{F}) = 0$.

Proof. We use the same notations as in (3.3).

(a) If $q < n - d$, then we have the exact sequence

$$0 \to H_A^q(U, \mathcal{F}) \to H_A^{q+1}(U, \mathcal{K})$$

and, since the coboundary map is continuous, we get

$$N_A^q(U, \mathcal{F}) \subset N_A^{q+1}(U, \mathcal{K}).$$

Hence to prove the lemma we can assume that $q = n - d$.

(b) $q = n - d$. Let $\mathcal{B}_j$, $\mathcal{Z}_j$, and $\mathcal{L}_j$ be the same as in (3.3). Using essentially the same induction proof as in (3.3), we obtain an

ℓ such that

$$\Gamma(U, \mathcal{J}_q)^{\ell}\, H_A^{q-1}(U, \mathcal{B}_1) = 0$$

for any open subset U of X, where ℓ is independent of U. Then from the exact sequence

$$0 \to H_A^{q-1}(U, \mathcal{B}_1) \xrightarrow{\delta} H_A^q(U, \mathcal{Z}_0) \xrightarrow{\alpha} H_A^q(U, {}_n\mathcal{O}^{s_0})$$

we obtain

$$\alpha(N_A^q(U, \mathcal{Z}_0)) = 0$$

by Lemma (4.2). Hence

$$N_A^q(U, \mathcal{Z}_0) \subset \operatorname{Im} \delta$$

and so is annihilated by $\Gamma(U, \mathcal{J}_q)^{\ell}$. By using the same argument and observing that the support of $\mathcal{L}_i$ is contained in the subvariety $S_{n-1}(\mathcal{F})$ whose ideal-sheaf is $\mathcal{J}_q$, we obtain the lemma from the exact sequences

$$\cdots \to H_A^{q-1}(U, \mathcal{L}_0) \to H_A^q(U, \mathcal{B}_0) \to H_A^q(U, \mathcal{Z}_0) \to \cdots$$

and

$$\cdots \to H_A^q(U, \mathcal{L}_{-1}) \to H_A^q(U, \mathcal{F}) \to H_A^q(U, \mathcal{B}_0) \to \cdots \quad .$$

Q.E.D.

(4.4) <u>Lemma</u>. Let $D \subset \mathbb{C}^n$ be a domain, V a subvariety of D, and $\mathcal{F}$ a coherent analytic sheaf on D such that $\mathcal{H}_V^0\mathcal{F} = 0$. Let $\mathcal{I}$ be the ideal-sheaf of V. If K is a holomorphically convex compact subset of D, then there exist an open neighborhood Y of K in D and sections $f_1,\dots,f_p \in \Gamma(Y,\mathcal{I})$ such that $(f_i)_x$ is not a zero-divisor of $\mathcal{F}_x$ for $x \in Y$ $(1 \leqq i \leqq p)$ and, for some $m \geqq 1$, $\mathcal{I}^m|K$ is contained in the ideal-sheaf generated by $f_1,\dots,f_p$ on K.

Proof. Since K is a holomorphically convex compact subset of D, there exist an open neighborhood Y of K in D and sections $g_1,\dots,g_q \in \Gamma(Y,\mathcal{I})$ such that $g_1,\dots,g_q$ generate $\mathcal{I}|Y$.

Since $\mathcal{H}_V^0\mathcal{F} = 0$, by Theorem (1.14)

$$\dim V \cap S_{k+1}(\mathcal{F}) \leqq k$$

for every k. Let S_{k+1} be the union of all $(k+1)$-dimensional branches of $Y \cap S_{k+1}(\mathcal{F})$. Choose a countable dense set $A = \{x_\nu\}$ in $\bigcup_k S_{k+1} - V$ such that $A \cap S_{k+1}$ is dense in $S_{k+1}-V$. For each x_ν,

$$(g_1(x_\nu),\dots,g_q(x_\nu)) \neq 0 ,$$

otherwise $x_\nu \in V$. Hence in $\mathbb{C}^q$ with coordinates $w_1,\dots,w_q$ we can find a linear form $\Sigma_{i=1}^q a_i w_i$ $(a_i \in \mathbb{C})$ such that

$$\Sigma_{i=1}^q a_i g_i(x_\nu) \neq 0$$

for each ν. Let

$$f_1 = \Sigma_{i=1}^{q} a_i g_i \in \Gamma(Y, \mathcal{I}).$$

Since $f_1(x_\nu) \neq 0$ for each ν,

$$\dim V(f_1) \cap S_{k+1}(\mathcal{F}) \leqq k$$

for every k, where $V(f_1)$ is the subvariety of Y defined by f_1. By Corollary (1.18), $(f_1)_x$ is not a zero-divisor of $\mathcal{F}_x$ for $x \in Y$. By induction and the same argument we can find

$$f_1, \ldots, f_p \in \Gamma(Y, \mathcal{I})$$

such that f_i does not vanish at any point of some countable subset B of

$$(\cup_k S_{k+1}) \cup (\cap_{k=1}^{i-1} V(f_k)) - V$$

whose intersection with S_{k+1} is dense in $S_{k+1} - V$ and whose intersection with every branch W of $\cap_{k=1}^{i-1} V(f_k)$ is dense in W - V. By Corollary (1.18), $(f_i)_x$ is not a zero-divisor of $\mathcal{F}_x$ for $x \in Y$. When $p \geqq n + 1$, $V = \cap_{i=1}^{p} V(f_i)$. The lemma follows from Hilbert Nullstellensatz. Q.E.D.

(4.5) Lemma. Let $D \subset \mathbb{C}^n$ be a domain, $A \subset V$ two subvarieties of D, and $\mathcal{F}$ a coherent analytic sheaf on D. Suppose $q \geqq 1$. Let S'_{k+q} be the union of all branches of $S_{k+q}(\mathcal{F})$ which are not contained in V. Assume that $\dim A \cap S'_{k+q} \leqq k$ for every k. Let $\mathcal{I}$ be the ideal-sheaf of V and let X be a relatively compact open subset of D whose topological closure X^- is holomorphically convex. Then there exists an integer $\ell \geqq 0$ such that, for any Stein open subset U of X, $\Gamma(U,\mathcal{I})^{\ell} N^q_A(U,\mathcal{F}) = 0$.

Proof. We may assume that $\mathcal{H}^0_V\mathcal{F} = 0$. For, by setting $\mathcal{G} = \mathcal{F}/\mathcal{H}^0_V\mathcal{F}$, we have the exact sequence

$$\cdots \to H^q_A(U, \mathcal{H}^0_V\mathcal{F}) \to H^q_A(U,\mathcal{F}) \to H^q_A(U,\mathcal{G}) \to \cdots .$$

Since, for some integer $\ell \geqq 0$, $\mathcal{I}^{\ell}\,\mathcal{H}^0_V\mathcal{F} = 0$ on some neighborhood of X^-, we have

$$\Gamma(U,\mathcal{I})^{\ell}\, H^q_A(U,\mathcal{H}^0_V\mathcal{F}) = 0$$

and the result follows from the corresponding result for $\mathcal{G}$ which satisfies $\mathcal{H}^0_V\mathcal{G} = 0$.

As in the proof of (3.4) let $A_k = A \cap S'_{k+q}$ and $A'_{k+1} = A_{k+1} - S'_{k+q}$. Since $(X - S'_{q+k})^- = X^-$ and $S_{q+k}(\mathcal{F}) - S'_{q+k} \subset V$, by Lemma (4.3) we have

$$\Gamma(U - S'_{k+q}, \mathcal{I})^{\ell}\, N^q_{A_{k+1}}(U - S'_{k+q}, \mathcal{F}) = 0,$$

where ℓ is independent of U. Now we have the following commutative diagram with an exact sequence

$$\begin{array}{ccccccc} H^{q-1}_{A'_{k+1}}(U, \mathfrak{F}) & \xrightarrow{\delta} & H^q_{A_k}(U, \mathfrak{F}) & \xrightarrow{\alpha} & H^q_{A_{k+1}}(U, \mathfrak{F}) & \rightarrow & H^q_{A'_{k+1}}(U, \mathfrak{F}) \\ & & & & & \overset{\beta}{\searrow} & \| \\ & & & & & & H^q_{A_{k+1}}(U-S'_{k+q}, \mathfrak{F}) . \end{array}$$

Since β is continuous, we obtain

$$\Gamma(U, \mathcal{J})^{\ell}\, N^q_{A_{k+1}}(U, \mathfrak{F}) \subset \operatorname{Im} \alpha .$$

By Lemma (3.3) we have also

$$\Gamma(U, \mathcal{J})^{\ell}\, H^{q-1}_{A'_{k+1}}(U, \mathfrak{F}) = 0$$

for ℓ large enough. As in the proof of Lemma (3.4) we are going to prove by induction on k that for ℓ large enough

$$\Gamma(U, \mathcal{J})^{\ell}\, N^q_{A_k}(U, \mathfrak{F}) = 0.$$

For k = 0, dim $A_0 \leqq 0$, and, since $S_{q-1}(\mathfrak{F}) \subset V$ in some open neighborhood of A_0, the result follows from Lemma (4.3). By Lemma (4.4) there exist an open neighborhood Y of X^- in D and sections

$$f_1, \ldots, f_p \in \Gamma(Y, \mathcal{J})$$

such that $(f_i)_x$ is not a zero-divisor of $\mathcal{F}_x$ for $x \in Y$ ($1 \leqq i \leqq p$) and, for some $m \geqq 1$, $\mathcal{I}^m | X^-$ is contained in the ideal-sheaf $\mathcal{J}$ generated by $f_1, \dots, f_p$ on X^-. For our purpose we may assume that $\mathcal{I} = \mathcal{J}$ on X, because annihilation by some power of $\Gamma(U, \mathcal{I})$ is equivalent to annihilated by some power of $\Gamma(U, \mathcal{J})$ for $U \subset X$.

Since U is Stein, we have the isomorphisms

$$H^q_{A_k}(U, \mathcal{F}) = \Gamma(U-A_k, \mathcal{F})/\Gamma(U, \mathcal{F}) \quad \text{for } q = 1$$

and

$$H^q_{A_k}(U, \mathcal{F}) = H^{q-1}(U-A_k, \mathcal{F}) \quad \text{for } q \geqq 2.$$

Let $\mathcal{U}$ be a Stein covering of $U - A_k$ and $\mathcal{V}$ be a Stein covering of $U - A_{k+1}$ which refines the restriction of $\mathcal{U}$ to $U - A_{k+1}$. α is induced by the map

$$\theta : Z^{q-1}(\mathcal{U}, \mathcal{F}) \oplus C^{q-2}(\mathcal{V}, \mathcal{F}) \to Z^{q-1}(\mathcal{V}, \mathcal{F})$$

defined by

$$\theta(\xi \oplus \eta) = \xi | V + \delta\eta ,$$

where, when $q = 1$, $C^{-1}(\mathcal{V}, \mathcal{F}) = \Gamma(U, \mathcal{F})$ and $\delta : C^{-1}(\mathcal{V}, \mathcal{F}) \to Z^0(\mathcal{V}, \mathcal{F})$ is the restriction map $\Gamma(U, \mathcal{F}) \to \Gamma(U-A_{k+1}, \mathcal{F})$. Since

$$\Gamma(U, \mathcal{I})^\ell N^q_{A_{k+1}}(U, \mathcal{F}) \subset \text{Im } \alpha .$$

we have

$$g\,\overline{B}^{q-1}(\mathcal{V}, \mathcal{F}) \subset \operatorname{Im}\theta$$

for any $g \in \Gamma(U, \mathcal{J})^{\ell}$, where $\overline{B}^{q-1}(\mathcal{V}, \mathcal{F})$ is the topological closure of $B^{q-1}(\mathcal{V}, \mathcal{F})$ in $Z^{q-1}(\mathcal{V}, \mathcal{F})$. Let now $g = f_{i_1} \ldots f_{i_\ell}$. Then g_x is not a zero-divisor of $\mathcal{F}_x$ for $x \in X$ and hence the map

$$C^{q-1}(\mathcal{V}, \mathcal{F}) \to C^{q-1}(\mathcal{V}, \mathcal{F})$$

defined by multiplication by g is a topological monomorphism. Consequently $g\,\overline{B}^{q-1}(\mathcal{V}, \mathcal{F})$ is a closed subspace and hence Fréchet. By the open mapping theorem for Fréchet spaces, the map

$$\theta^{-1}(g\,\overline{B}^{q-1}(\mathcal{V}, \mathcal{F})) \to g\,\overline{B}^{q-1}(\mathcal{V}, \mathcal{F})$$

induced by θ is open. If $\xi \in \overline{B}^{q-1}(\mathcal{V}, \mathcal{F})$ and $\{\xi_\nu\}$ is a sequence in $B^{q-1}(\mathcal{V}, \mathcal{F})$ converging to ξ, then we can find a sequence $\{\zeta_\nu \oplus \eta_\nu\}$ in $\theta^{-1}(g\,\overline{B}^{q-1}(\mathcal{V}, \mathcal{F}))$ converging to some $\zeta \oplus \eta$ such that

$$\theta(\zeta_\nu \oplus \eta_\nu) = g\,\xi_\nu \in B^{q-1}(\mathcal{V}, \mathcal{F})$$

and $\theta(\zeta \oplus \eta) = g\xi$. Since $g\,\xi_\nu \in B^{q-1}(\mathcal{V}, \mathcal{F})$, we know that $h\zeta_\nu \in B^{q-1}(\mathfrak{U}, \mathcal{F})$ for each $h \in \Gamma(U, \mathcal{J})^{\ell}$, because the class in $H^q_{A_k}(U, \mathcal{F})$ defined by ζ_ν is in the image of δ. Hence $h\zeta \in \overline{B}^{q-1}(\mathfrak{U}, \mathcal{F})$

and by induction hypothesis for $h_1 \in \Gamma(U, \mathcal{J})^{\ell_1}$ and ℓ_1 large enough

$$h_1 h\zeta \in B^{q-1}(\mathfrak{U}, \mathcal{F}).$$

Hence we have

$$h_1 hg\xi \in B^{q-1}(\mathcal{V}, \mathcal{F}).$$

Since $f_1|U, \ldots, f_p|U$ generate $\Gamma(U, \mathcal{J})$ for every $U \subset X$, it follows that

$$\Gamma(U, \mathcal{J})^{\ell_1 + 2\ell} N^q_{A_{k+1}}(U, \mathcal{F}) = 0$$

if U is Stein. Hence the induction on k is complete. Q.E.D.

(4.6) <u>Proposition</u>. Let D be a domain in $\mathbb{C}^n$, $A \subset D$ a subvariety, and $\mathcal{F}$ a coherent analytic sheaf on D such that $\dim A \cap \overline{S}_{k+q}(\mathcal{F}|D-A) \leqq k$ for every k. Then, for any relatively compact Stein domain $X \subset\subset D$ whose topological closure is holomorphically convex, $\mathcal{N}^i_A \mathcal{F} = 0$ for $i \leqq q$ and $H^i_A(X, \mathcal{F})$ is a Fréchet space for $i \leqq q$. (Note that $\mathcal{H}^q_A \mathcal{F}$ is no longer coherent in general.)

Proof. Let $\mathcal{J}$ be the ideal-sheaf of A. We are going to prove by induction on q that for every Stein open subset U of X the space $H^i_A(U, \mathcal{F})$ is Hausdorff for $i \leqq q$ and that for any $\Gamma(U, {}_n\mathcal{O})$-homomorphism

$$\varphi: \Gamma(U, {}_n\mathcal{O})^r \to H^q_A(U, \mathcal{F})$$

if Im φ is annihilated by some power of $\Gamma(U, \mathcal{J})$, then Im φ is closed.

For $q = 0$ what is to be proved is trivial. Let $q \geqq 1$ and assume that U is a Stein open subset of X. By Lemma (3.4) and (4.5) there exists an integer $\ell \geqq 0$ such that

$$\Gamma(U, \mathcal{J})^{\ell}\, H_A^i(U, \mathcal{F}) = 0$$

for $i < q$ and

$$\Gamma(U, \mathcal{J})^{\ell}\, N_A^q(U, \mathcal{F}) = 0\ .$$

By Lemma (4.4) we can choose an open neighborhood Y of X^- in D and $g \in \Gamma(Y, \mathcal{J})^{\ell}$ such that g_x is not a zero-divisor of $\mathcal{F}_x$ for $x \in Y$. We obtain the exact sequence

$$(*) \qquad H_A^{q-1}(U, \mathcal{F}) \xrightarrow{\alpha} H_A^{q-1}(U, \mathcal{F}/g\mathcal{F}) \xrightarrow{\delta} H_A^q(U, \mathcal{F}) \xrightarrow{g^*} H_A^q(U, \mathcal{F}).$$

By induction hypothesis $H_A^{q-1}(U, \mathcal{F}/g\mathcal{F})$ is a Fréchet space. Since $\mathcal{H}_A^{q-1}\mathcal{F}$ is coherent, we have a sheaf-epimorphism $\psi\colon {}_n\mathcal{O}^s \to \mathcal{H}_A^{q-1}\mathcal{F}$ on X. Since

$$H_A^{q-1}(U, \mathcal{F}) = \Gamma(U, \mathcal{H}_A^{q-1}\mathcal{F}),$$

ψ induces a $\Gamma(U, {}_n\mathcal{O})$-epimorphism

$$\psi^*\colon \Gamma(U, {}_n\mathcal{O})^s \to H_A^{q-1}(U, \mathcal{F}).$$

Since Im $\alpha\psi^* = \text{Im}\ \alpha$ is annihilated by $\Gamma(U, \mathcal{J})^{\ell}$, by induction hypothesis Im α is closed. Since $g^* N_A^q(U, \mathcal{F}) = 0$, we have $N_A^q(U, \mathcal{F}) \subset \text{Im}\ \delta$. By considering Čech cohomology, we conclude easily that the map

$$\delta^{-1}(N_A^q(U, \mathcal{F})) \to N_A^q(U, \mathcal{F})$$

is open. Hence $N_A^q(U, \mathcal{F})$ is isomorphic to the Fréchet space

$$\delta^{-1}(N_A^q(U, \mathcal{F}))/\text{Im}\ \alpha\ .$$

It follows that $N_A^q(U, \mathcal{F}) = 0$ and $H_A^q(U, \mathcal{F})$ is Hausdorff. That $H_A^i(U, \mathcal{F})$ is Hausdorff for $i < q$ is part of the induction hypothesis.

Suppose φ: $\Gamma(U, {}_n\mathcal{O})^r \to H_A^q(U, \mathcal{F})$ is a $\Gamma(U, {}_n\mathcal{O})$-homomorphism such that Im φ is annihilated by some power of $\Gamma(U, \mathcal{J})$. We may assume without loss of generality that $\Gamma(U, \mathcal{J})^{\ell}\ \text{Im}\ \varphi = 0$, where ℓ is the same as before. From (*) we have Im $\varphi \subset$ Im δ. There exists a $\Gamma(U, {}_n\mathcal{O})$-homomorphism

$$\tilde{\varphi}: \Gamma(U, {}_n\mathcal{O})^r \to H_A^{q-1}(U, \mathcal{F}/g\mathcal{F})$$

such that $\varphi = \delta\tilde{\varphi}$. Define

$$\gamma: \Gamma(U, {}_n\mathcal{O})^s \oplus \Gamma(U, {}_n\mathcal{O})^r \dot{\to} H_A^{q-1}(U, \mathcal{F}/g\mathcal{F})$$

by

$$\gamma(a \oplus b) = \alpha\psi^*(a) + \tilde{\varphi}(b).$$

Since $\Gamma(U, \mathcal{J})^{\ell}\ \delta(\mathrm{Im}\ \tilde{\varphi}) = 0$, Im γ is annihilated also by some power of $\Gamma(U, \mathcal{J})$ and therefore by induction hypothesis is closed. Ker δ is closed, because $H^q_A(U, \mathfrak{F})$ is Hausdorff. Since $\delta^{-1}(\mathrm{Im}\ \varphi) = \mathrm{Im}\ \gamma$, from the topological isomorphism $\mathrm{Im}\ \varphi \approx \mathrm{Im}\ \gamma/\mathrm{Ker}\ \delta$ we conclude that Im φ is closed. The induction is complete.

By setting $U = X$, we conclude that $H^i_A(X, \mathfrak{F})$ is Hausdorff for $i \leq q$. Being the open continuous image of a Fréchet space, $H^i_A(X, \mathfrak{F})$ is Fréchet for $i \leq q$. By letting U run through a neighborhood basis of Stein open sets, we obtain $\mathcal{N}^i_A \mathfrak{F} = 0$ for $i \leq q$.

Q.E.D.

(4.7) Theorem. Let X be a complex space, $A \subset X$ a subvariety, and $\mathfrak{F}$ a coherent analytic sheaf on X such that $\mathcal{H}^i_A \mathfrak{F}$ is coherent for $i < q$. Then for every Stein open subset $\Omega \subset X$, $H^i_A(\Omega, \mathfrak{F})$ is a Fréchet space for $i \leq q$.

Proof. Because of the canonical isomorphism

$$H^i_A(\Omega, \mathfrak{F}) \to \Gamma(\Omega, \mathcal{H}^i_A \mathfrak{F}) \qquad (i \leq q),$$

the map

$$N^i_A(\Omega, \mathfrak{F}) \to \Gamma(\Omega, \mathcal{N}^i_A \mathfrak{F}) \qquad (i \leq q)$$

is injective and hence it suffices to show that $\mathcal{N}^i_A \mathfrak{F} = 0$ for $i \leq q$. This follows from Proposition (4.6), because the topology of $H^i_A(U, \mathfrak{F})$

does not depend on the imbedding of U into a complex number space.

Q.E.D.

(4.8) Corollary. The canonical isomorphism $H^i_A(\Omega, \mathcal{F}) \to \Gamma(\Omega, \mathcal{H}^i_A \mathcal{F})$ is a homeomorphism for $i < q$, where $\Gamma(\Omega, \mathcal{H}^i_A \mathcal{F})$ is given the usual Fréchet structure as the section module of a coherent analytic sheaf.

Proof. By the open mapping theorem for Fréchet spaces, it suffices to show that the map

$$\lambda^i\colon \Gamma(\Omega, \mathcal{H}^i_A \mathcal{F}) \to H^i_A(\Omega, \mathcal{F})$$

is continuous. By the closed graph theorem, it suffices to show that λ^i has a closed graph. Let $s_\nu \to s$ in $\Gamma(\Omega, \mathcal{H}^i_A \mathcal{F})$ and let $\xi_\nu \to \xi$ in $H^i_A(\Omega, \mathcal{F})$ such that $\xi_\nu = \lambda^i(s_\nu)$. If $\Omega' \subset\subset \Omega$ is Stein, then $\mathcal{H}^i_A \mathcal{F}$ is generated by a finite number of sections

$$t_1, \ldots, t_m \in \Gamma(\Omega', \mathcal{H}^i_A \mathcal{F}).$$

Since $s_\nu | \Omega' \to s | \Omega'$, there are holomorphic functions $f^{(\nu)}_\mu$ on Ω' converging to some holomorphic function f_μ on Ω' such that

$$s_\nu | \Omega' = \Sigma^m_{\mu=1} f^{(\nu)}_\mu t_\mu .$$

Let $\zeta_\mu \in H^i_A(\Omega', \mathcal{F})$ be the image of t_μ under the canonical isomorphism

$$\lambda^i_{\Omega'}: \Gamma(\Omega', \mathcal{H}^i_A \mathfrak{F}) \to H^i_A(\Omega', \mathfrak{F}).$$

Then

$$\xi_\nu|\Omega' = \lambda^i_{\Omega'}(s_\nu|\Omega') = \Sigma^m_{\mu=1} f^{(\nu)}_\mu \zeta_\mu ,$$

where $\xi_\nu|\Omega'$ is the image of ξ_ν under the map

$$H^i_A(\Omega, \mathfrak{F}) \to H^i_A(\Omega', \mathfrak{F}).$$

Hence

$$\xi|\Omega' = \Sigma_{\mu=1} f_\mu \zeta_\mu = \lambda^i_{\Omega'}(s|\Omega') = \lambda^i(s)|\Omega' .$$

Since $H^i_A(\Omega, \mathfrak{F})$ is algebraically a section module, we conclude that $\xi = \lambda^i(s)$ and hence the graph of λ^i is closed for $i < q$. Q.E.D.

§5 Duality

When the exceptional set degenerates to a point, the theory of analytic sheaves of local cohomology discussed earlier can also be derived by the techniques of Serre duality. Since the case of an exceptional point is very similar to the more general case of an holomorphically convex compact subset, in what follows we will treat the latter case. For obvious reasons we cannot prove all preparatory theorems we need. Those we cannot prove we refer to the following three sources: [6], [16], and [21].

If X is a topological space and $\mathcal{F}$ is a sheaf of abelian groups, we denote by $H_c^i(X, \mathcal{F})$ the cohomology groups of $\mathcal{F}$ with compact supports. They are defined in the following way. If

$$0 \to \mathcal{F} \to \mathcal{C}^0 \to \mathcal{C}^1 \to \dots$$

is a flabby resolution of $\mathcal{F}$, then we have the complex

$$0 \to \Gamma_c(X, \mathcal{C}^0) \to \Gamma_c(X, \mathcal{C}^1) \to \dots \quad ,$$

where $\Gamma_c(X, \mathcal{C}^i)$ is the submodule of all sections of $\Gamma(X, \mathcal{C}^i)$ having compact supports. The cohomology groups of the complex are $H_c^i(X, \mathcal{F})$.

(5.1) Let X be a topological space, $K \subset X$ a compact subset, and $\mathcal{F}$ a sheaf of abelian groups on X. Then there is a canonical exact sequence

$$0 \to H_c^0(X-K, \mathcal{F}) \to H_c^0(X, \mathcal{F}) \to H^0(K, \mathcal{F}) \to H_c^1(X-K, \mathcal{F}) \to H_c^1(X, \mathcal{F}) \to \dots$$

[6, Th. (4.10.1)].

(5.2) Let X be a real C^∞ manifold of dimension m. Let $\mathcal{E}^m$ be the sheaf of (complex-valued) differentiable forms of degree m. For an open subset U of X let $\Gamma_c(U, \mathcal{E}^m)$ be the space of all differentiable forms of degree m on U with compact supports. $\Gamma_c(U, \mathcal{E}^m)$ is given the inductive limit topology

$$\varinjlim \{\Gamma_K(U, \mathcal{E}^m) \mid K \text{ is a compact subset of } U\},$$

where $\Gamma_K(U, \mathcal{E}^m)$ is the set of all differentiable forms of degree m on U with supports contained in K and is given the topology of uniform convergence of the coefficients of the forms and their derivatives. Let $\mathcal{K}(U)$ be the topological dual of $\Gamma_c(U, \mathcal{E}^m)$. If $V \subset U$ is an open subset, we obtain a natural restriction map $\mathcal{K}(U) \to \mathcal{K}(V)$ from the inclusion

$$\Gamma_c(V, \mathcal{E}^m) \to \Gamma_c(U, \mathcal{E}^m).$$

The presheaf $U \mapsto \mathcal{K}(U)$ defines the sheaf $\mathcal{K}$ of germs of distribution m-forms on X.

(5.3) Let $(X, \mathcal{O})$ be a complex manifold of dimension n. Let

Ω^r = the sheaf of germs of holomorphic r-forms on X,

$\mathcal{A}^{p,q}$ = the sheaf of germs of C^∞ (p,q)-forms on X,

$\mathcal{K}^{p,q}$ = the sheaf of germs of distribution (p,q)-forms on X.

By a well-known theorem of Dolbeault [3] we obtain by $\bar{\partial}$ differentiation the exact sequences

$$0 \to \Omega^p \to \mathcal{A}^{p,0} \to \mathcal{A}^{p,1} \to \ldots \to \mathcal{A}^{p,n} \to 0 ,$$

$$0 \to \Omega^p \to \mathcal{K}^{p,0} \to \mathcal{K}^{p,1} \to \ldots \to \mathcal{K}^{p,n} \to 0 .$$

The sheaves $\mathcal{A}^{p,q}$ and $\mathcal{K}^{p,q}$ are soft sheaves and are natural $\mathcal{O}$-sheaves. If $\mathcal{F}$ is a locally free $\mathcal{O}$-sheaf on X, then $\mathcal{F} \otimes_{\mathcal{O}} \mathcal{A}^{p,q}$ and $\mathcal{F} \otimes_{\mathcal{O}} \mathcal{K}^{p,q}$ are soft and

$$0 \to \mathcal{F} \otimes_{\mathcal{O}} \Omega^p \to \mathcal{F} \otimes_{\mathcal{O}} \mathcal{A}^{p,0} \to \ldots \to \mathcal{F} \otimes_{\mathcal{O}} \mathcal{A}^{p,n} \to 0 ,$$

$$0 \to \mathcal{F} \otimes_{\mathcal{O}} \Omega^p \to \mathcal{F} \otimes_{\mathcal{O}} \mathcal{K}^{p,0} \to \ldots \to \mathcal{F} \otimes_{\mathcal{O}} \mathcal{K}^{p,n} \to 0$$

are soft resolutions of $\mathcal{F} \otimes_{\mathcal{O}} \Omega^p$. For any open subset Y of X let

$$A^{p,q}(Y) = \Gamma(Y, \mathcal{F}^* \otimes_{\mathcal{O}} \mathcal{A}^{p,q}),$$

$$K_c^{p,q}(Y) = \Gamma_c(Y, \mathcal{F} \otimes_{\mathcal{O}} \mathcal{K}^{p,q}),$$

where $\mathcal{F}^* = \mathrm{Hom}_{\mathcal{O}}(\mathcal{F}, \mathcal{O})$. Since $\mathcal{F}$ and $\mathcal{F}^*$ are locally free, on some open neighborhood of every point $\mathcal{F}^* \otimes_{\mathcal{O}} \mathcal{A}^{p,q}$ is a direct sum of a finite number of copies of the sheaf $\mathcal{E}$ of germs of differentiable functions. Hence $A^{p,q}(Y)$ can be equipped with the topology of local uniform convergence of such function tuples and their derivatives. $A^{p,q}(Y)$ is an (FS)-space (Fréchet-Schwartz space [13, p. 277]). The following two statements are proved in [21].

(5.4) The topological dual of $A^{p,q}(Y)$ is the space $K_c^{n-p,n-q}(Y)$.

In the following $K_c^{n-p,n-q}(Y)$ will always have the topology of the strong dual of $A^{p,q}(Y)$.

(5.5) The transpose of the continuous linear map $\overline{\partial}: A^{p,q}(Y) \to A^{p,q+1}(Y)$ is the map $(-1)^{p+q+1}\overline{\partial}: K_c^{n-p,n-q-1}(Y) \to K_c^{n-p,n-q}(Y)$.

We will need the following two lemmas on topological vector spaces.

(5.6) Lemma. Let E and F be both (FS)-spaces or both (DFS)-spaces (strong duals of (FS)-spaces). If u: E $\to$ F is a continuous linear map with closed range, then u is a topological homomorphism and its transpose u^t: F' $\to$ E' is also a topological homomorphism with closed range, where E' and F' are strong duals of E and F respectively.

Proof. In the case of (FS)-spaces the usual open mapping theorem for Frechet spaces implies that u is a topological homomorphism.

Suppose E and F are (DFS)-spaces and Im u is closed. Since $(\text{Im } u)^{\perp\perp}$ = Im u, by [13, p. 285, Cor. to Prop. 10] Im u is the strong dual of $F'/(\text{Im } u)^{\perp}$. By [13, p. 279, Cor. to Prop. 7] $F'/(\text{Im } u)^{\perp}$ is an (FS)-space and is therefore Montel [13, p. 277, Cor. to Prop. 4] and reflexive [13, p. 231, Cor. to Prop. 1]. Consequently Im u is reflexive and hence is barrelled [13, p. 229, Cor. to Prop. 6]. Since E' is an (FS)-space and hence is reflexive, by [13, p. 300, Prop. 6] E is Pták. By [13, p. 300, Prop. 5] E/Ker u is Pták. Since the inverse map v: Im u $\to$ E/Ker u induced by u has a closed graph, by the generalized closed graph theorem [13, p. 301, Th. 4] v is continuous. It follows that u is a topological homomorphism.

In both cases the closedness of Im u^t follows canonically from the fact that u is a topological homomorphism with closed range. From the previous arguments u^t is also a topological homomorphism, because only (FS)-spaces or (DFS)-spaces are involved. Q.E.D.

(5.7) Proposition. Let $E \xrightarrow{u} F \xrightarrow{v} G$ be a complex of (FS)-spaces or (DFS)-spaces and continuous linear maps. Let $G' \xrightarrow{v^t} F' \xrightarrow{u^t} E'$ be the dual complex. Let H = Ker v/Im u and H* = Ker u^t/im v^t. If Im u and Im v are closed, then u, v, u^t, v^t are topological homomorphisms with closed ranges and H* equals the strong dual H' of H both algebraically and topologically.

Proof. By Lemma (5.6) u, v, u^t, v^t are topological homomorphisms with closed ranges. To prove the last statement, first let E, F, G be (FS)-spaces. Define the canonical map κ: H' $\rightarrow$ H* as follows. If $h \in H'$, let $\tilde{h}$ be a Hahn-Banach extension of $h \circ p$ to F, where p: Ker v $\rightarrow$ H is the projection. Then

$$u^t(\tilde{h}) = \tilde{h} \circ u = h \circ p \circ u = 0.$$

Hence $\tilde{h} \in$ Ker u^t defines an element $\kappa(h) \in H^*$. Since u, v are homomorphisms with closed range, it can easily be verified that the definition of κ does not depend on the choice of $\tilde{h}$ and that κ is an algebraic isomorphism. κ is continuous. For, if $\Omega \subset$ Ker u^t is a neighborhood of zero, then $\Omega \supset$ Ker $u^t \cap B^0$ for some bounded subset B of F (where B^0 is the polar of B) and $(p(B \cap \text{Ker } v))^0$ is a neighborhood of 0 in H' which is mapped by κ into $p^*\Omega$ in H*, where p^*: Ker $u^t \rightarrow H^*$ is the projection. Since H* is barrelled and as a (DFS)-space H' is Pták, by [13, p. 299, Prop. 2] κ is open and hence a topological isomorphism.

If now E, F, G are (DFS)-spaces, we get in the same way a continuous linear isomorphism $\kappa: H' \to H^*$. Since

$$E \xrightarrow{u} F \xrightarrow{v} G$$

is the dual complex of

$$G' \xrightarrow{{}^t u} F' \xrightarrow{{}^t v} G',$$

by the previous argument $(H^*)'$ is topologically isomorphic to H. Hence H is a (DFS)-space. Since H' is an (FS)-space, by the open mapping theorem for Fréchet spaces, κ is a topological isomorphism.

Q.E.D.

(5.8) Theorem. Let X be a Stein manifold of dimension n and $\mathfrak{F}$ be a locally free analytic sheaf on X. Then $H_c^q(X, \mathfrak{F} \otimes_{\mathcal{O}} \Omega^p) = 0$ for $q \neq n$ and $H_c^n(X, \mathfrak{F} \otimes_{\mathcal{O}} \Omega^p)$ is the topological dual of the (FS)-space $H^0(X, \mathfrak{F}^* \otimes_{\mathcal{O}} \Omega^{n-p})$.

Proof. Consider the complexes

$$0 \to K_c^{p,0}(X) \to K_c^{p,1}(X) \to \dots \to K^{p,n}(X) \to 0,$$

$$0 \leftarrow A^{n-p,n}(X) \leftarrow A^{n-p,n-1}(X) \leftarrow \dots \xleftarrow{\bar{\partial}^0} A^{n-p,0}(X) \leftarrow 0.$$

Since X is Stein, by Cartan's theorem B the second complex is exact except that

$$\operatorname{Ker} \bar{\partial}^0 = H^0(X, \mathcal{F}^* \otimes_{\mathcal{O}} \Omega^{n-p}).$$

Hence by Lemma (5.6) all maps in both complexes are topological homomorphisms with closed ranges. By Proposition (5.7) the result follows. Q.E.D.

The following is the version of the preceding theorem for a fixed Stein compact set.

(5.9) <u>Theorem</u>. Let X be a Stein manifold of dimension n, $K \subset X$ be a Stein compact set (i.e. a compact set admitting a neighborhood basis of Stein open subsets). Then for a locally free analytic sheaf $\mathcal{F}$ on X,

(i) $H^q_K(X, \mathcal{F} \otimes_{\mathcal{O}} \Omega^p) = 0$ for $q \neq n$,

(ii) $H^0(K, \mathcal{F}^* \otimes_{\mathcal{O}} \Omega^{n-p})$ with the natural inductive limit topology is a (DFS)-space,

(iii) $H^n_K(X, \mathcal{F} \otimes_{\mathcal{O}} \Omega^p)$ with its natural topology is the strong dual of $H^0(K, \mathcal{F}^* \otimes_{\mathcal{O}} \Omega^{n-p})$.

Proof. (a) First we are going to prove that $H^{n-1}(X-K, \mathcal{F} \otimes_{\mathcal{O}} \Omega^p)$ is an (FS)-space and $H^1_c(X-K, \mathcal{F}^* \otimes_{\mathcal{O}} \Omega^{n-p})$ is its strong dual.

Let $Y = X - K$. We have the complexes

$$K_c^{n-p,0}(Y) \xrightarrow{\bar{\partial}^0} K_c^{n-p,1}(Y) \xrightarrow{\bar{\partial}^1} K_c^{n-p,2}(Y),$$

$$A^{p,n}(Y) \xleftarrow{\bar{\partial}^{n-1}} A^{p,n-1}(Y) \xleftarrow{\bar{\partial}^{n-2}} A^{p,n-2}(Y).$$

Since by Theorem (5.8) $H_c^2(Y, \mathfrak{F}^* \otimes_{\mathcal{O}} \Omega^{n-p})$ is either 0 or Hausdorff, by Lemma (5.6) $\overline{\partial}^1$ and its transpose $\overline{\partial}^{n-2}$ are topological homomorphisms with closed ranges. Since Y has no compact component and $\mathfrak{F} \otimes_{\mathcal{O}} \Omega^p$ is locally free, by [14] $H^n(Y, \mathfrak{F} \otimes_{\mathcal{O}} \Omega^p) = 0$. Consequently $\overline{\partial}^{n-1}$ is surjective and hence a topological homomorphism with closed range. By Proposition (5.7) $H^{n-1}(X-K, \mathfrak{F} \otimes_{\mathcal{O}} \Omega^p)$ and $H_c^1(X-K, \mathfrak{F}^* \otimes_{\mathcal{O}} \Omega^{n-p})$ are respectively an (FS)-space and a (DFS)-space and are strong duals of each other.

(b) Since K is Stein, $H^q(K, \mathfrak{F}^* \otimes_{\mathcal{O}} \Omega^{n-p})$ equals

$$\varinjlim \{H^q(U, \mathfrak{F}^* \otimes_{\mathcal{O}} \Omega^{n-p}) \mid U \text{ is a Stein open neighborhood of } K\}$$

and hence by Cartan's theorem B is 0 for $q \geq 1$. From (5.1) and Theorem (5.8) we have

$$(*) \qquad H_c^q(X-K, \mathfrak{F}^* \otimes_{\mathcal{O}} \Omega^{n-p}) = 0 \text{ for } q \neq 1, n.$$

Moreover, when $n \geq 2$ the maps

$$\delta\colon H^0(K, \mathfrak{F}^* \otimes_{\mathcal{O}} \Omega^{n-p}) \to H_c^1(X-K, \mathfrak{F}^* \otimes_{\mathcal{O}} \Omega^{n-p})$$

$$\sigma\colon H_c^n(X-K, \mathfrak{F}^* \otimes_{\mathcal{O}} \Omega^{n-p}) \to H_c^n(X, \mathfrak{F}^* \otimes_{\mathcal{O}} \Omega^{n-p})$$

are algebraic isomorphisms. When n = 1, the sequence

$$0 \to H^0(K, \mathfrak{F}^* \otimes_{\mathcal{O}} \Omega^{1-p}) \xrightarrow{\delta} H_c^1(X-K, \mathfrak{F}^* \otimes_{\mathcal{O}} \Omega^{1-p}) \xrightarrow{\sigma} H_c^1(X, \mathfrak{F}^* \otimes_{\mathcal{O}} \Omega^{1-p}) \to 0$$

is exact.

We claim that δ is continuous. If $s \in H^0(K, \mathcal{F}^*\otimes_{\mathcal{O}} \Omega^{n-p})$, then $\delta(s)$ can be defined as follows (as can easily be verified through soft resolutions). We can find $T \in \Gamma(X, \mathcal{F}^*\otimes_{\mathcal{O}} \mathcal{K}^{n-p,0})$ which agrees with s on K. Then $\bar\partial T \in K_c^{n-p,1}(X-K)$, because, on some open neighborhood of K, T is locally some tuple of holomorphic functions. Since $\bar\partial(\bar\partial T) = 0$, $\bar\partial T$ defines an element in $H_c^1(X-K, \mathcal{F}^*\otimes_{\mathcal{O}} \Omega^{n-p})$ which is equal to $\delta(s)$. Since $\bar\partial$ is continuous, it is clear from this description of δ that δ is continuous.

Since $H_c^1(X-K, \mathcal{F}^*\otimes_{\mathcal{O}} \Omega^{n-p})$ is a (DFS)-space by (a) and δ is continuous, $H^0(K, \mathcal{F}^*\otimes_{\mathcal{O}} \Omega^{n-p})$ is Hausdorff. Being a closed subspace of the (DFS)-space $H_c^1(X-K, \mathcal{F}^*\otimes_{\mathcal{O}} \Omega^{n-p})$, Im δ is also a (DFS)-space (cf. the proof of Lemma (5.6)) and hence is barrelled. Since $H^0(K, \mathcal{F}^*\otimes_{\mathcal{O}} \Omega^{n-p})$ is Hausdorff and is the inductive limit of Fréchet (and hence Ptak) spaces, by applying the generalized closed graph theorem [13, p. 304, Prop. 10] to the inverse map

$$\eta: \operatorname{Im} \delta \to H^0(K, \mathcal{F}^*\otimes_{\mathcal{O}} \Omega^{n-p})$$

induced by δ, we conclude that η is continuous. Hence, when $n \geqq 2$, δ is a topological isomorphism; and when $n = 1$, δ maps $H^0(K, \mathcal{F}^*\otimes_{\mathcal{O}} \Omega^{n-p})$ homeomorphically onto Im δ.

σ is obviously continuous. In the case of $n = 1$, by applying Proposition (5.7) to the complex

$$0 \to H_c^1(X-K, \mathcal{F}^*\otimes_{\mathcal{O}} \Omega^{1-p}) \xrightarrow{\sigma} H_c^1(X, \mathcal{F}^*\otimes_{\mathcal{O}} \Omega^{1-p})$$

and its dual ((a) and Theorem (5.8))

$$0 \leftarrow H^0(X-K, \mathcal{F}\otimes_{\mathcal{O}} \Omega^p) \overset{\tau}{\leftarrow} H^0(X, \mathcal{F}\otimes_{\mathcal{O}} \Omega^p)$$

and by using the topological isomorphism

$$H^1_K(X, \mathcal{F}\otimes_{\mathcal{O}}\Omega^p) \approx \text{Coker } \tau,$$

we conclude that $H^1_K(X, \mathcal{F}\otimes_{\mathcal{O}} \Omega^p)$ is the strong dual of the (DFS)-space $H^0(K, \mathcal{F}^* \otimes_{\mathcal{O}} \Omega^{1-p})$. Hence the theorem is proved when $n = 1$. Since the theorem is trivially true when $n = 0$, in the rest of the proof we can assume that $n \geqq 2$.

Since σ is continuous and by Theorem (5.8) $H^n_c(X, \mathcal{F}\otimes_{\mathcal{O}} \Omega^{n-p})$ is (DFS), $H^n_c(X-K, \mathcal{F}^* \otimes_{\mathcal{O}} \Omega^{n-p})$ is Hausdorff. By applying Proposition (5.7) to the complex

$$K_c^{n-p,n-1}(X - K) \to K_c^{n-p,n}(X - K) \to 0$$

and its dual, we conclude that

(**) $\left\{ \begin{array}{l} H^n_c(X-K, \mathcal{F}^*\otimes_{\mathcal{O}} \Omega^{n-p}) \text{ is the strong dual of the (FS)-} \\ \text{space } H^0(X-K, \mathcal{F}\otimes_{\mathcal{O}} \Omega^p). \end{array} \right.$

By the generalized closed graph theorem [13, p. 301, Th. 4] applied to σ^{-1}, it follows that σ^{-1} is continuous. Therefore σ is a topological isomorphism.

(c) We are going to prove that $H^i_K(X, \mathcal{F}\otimes_{\mathcal{O}} \Omega^p) = 0$ for $0 \leqq i \leqq n - 1$.

Since X is Stein, we have the exact sequence

$$0 \to H_K^0(X, \mathcal{F}\otimes_{\mathcal{O}} \Omega^p) \to H^0(X, \mathcal{F}\otimes_{\mathcal{O}} \Omega^p) \xrightarrow{\rho} H^0(X-K, \mathcal{F}\otimes_{\mathcal{O}} \Omega^p) \to H_K^1(X, \mathcal{F}\otimes_{\mathcal{O}} \Omega^p) \to 0.$$

By Theorem (5.8) and (**) $H^0(X, \mathcal{F}\otimes_{\mathcal{O}} \Omega^p)$ and $H^0(X-K, \mathcal{F}\otimes_{\mathcal{O}} \Omega^p)$ are respectively the strong duals of the (DFS)-spaces $H_c^n(X, \mathcal{F}^*\otimes_{\mathcal{O}} \Omega^{n-p})$ and $H_c^n(X-K, \mathcal{F}^*\otimes_{\mathcal{O}} \Omega^{n-p})$. However, by (b)

$$\sigma\colon H_c^n(X-K, \mathcal{F}^*\otimes_{\mathcal{O}} \Omega^{n-p}) \to H_c^n(X, \mathcal{F}^*\otimes_{\mathcal{O}} \Omega^{n-p})$$

is a topological isomorphism. Hence ρ is an isomorphism. It follows that $H_K^i(X, \mathcal{F}\otimes_{\mathcal{O}} \Omega^p) = 0$ for $i = 0, 1$.

By (*) and Proposition (5.7), as in the proof of (**) we have

$$H^{n-q}(X-K, \mathcal{F}\otimes_{\mathcal{O}} \Omega^p) = 0$$

for $2 \leqq q \leqq n-1$. Hence, for $2 \leqq i \leqq n-1$,

$$H_K^i(X, \mathcal{F}\otimes_{\mathcal{O}} \Omega^p) = H^{i-1}(X-K, \mathcal{F}\otimes_{\mathcal{O}} \Omega^p) = 0.$$

(d) By (a), $H^{n-1}(X-K, \mathcal{F}^*\otimes_{\mathcal{O}} \Omega^{n-p})$ is the strong dual of the (DFS)-space $H_c^1(X-K, \mathcal{F}^* \otimes_{\mathcal{O}} \Omega^{n-p})$. However, by (b)

$$\delta\colon H^0(K, \mathcal{F}^*\otimes_{\mathcal{O}} \Omega^{n-p}) \to H_c^1(X-K, \mathcal{F}^*\otimes_{\mathcal{O}} \Omega^{n-p})$$

is a topological isomorphism; and it is always true that

$$H_K^n(X, \mathcal{F}\otimes_{\mathcal{O}} \Omega^p) = H^{n-1}(X-K, \mathcal{F}\otimes_{\mathcal{O}} \Omega^p).$$

Hence $H_K^n(X, \mathcal{F}\otimes_{\mathcal{O}} \Omega^p)$ is the strong dual of the (DFS)-space $H^0(K, \mathcal{F}^*\otimes_{\mathcal{O}} \Omega^{n-p})$. Q.E.D.

(5.10) Remark. If K is a compact polydisc or a point, then the proof of Theorem (5.9) is simpler, if we consider the duality between $H^0(K, \mathcal{O})$ and $H^{n-1}(X-K, \mathcal{O})$ given by the integral over the distinguished boundary.

(5.11) Remark. Since X is Stein, we have

$$H_c^q(X, \mathcal{F}) = \varinjlim\{H_K^q(X, \mathcal{F}) \mid K \text{ is a Stein compact subset of } X\}.$$

Hence Theorem (5.8) is a consequence of Theorem (5.9).

(5.12) Theorem. Let X be a Stein manifold of dimension n, $K \subset X$ a Stein compact set, and $\mathcal{F}$ a coherent analytic sheaf on X. Then

(i) $H^0(K, \mathcal{E}xt_{\mathcal{O}}^{n-q}(\mathcal{F}, \mathcal{O})\otimes_{\mathcal{O}} \Omega^{n-p})$ with its natural inductive limit topology is a (DFS)-space,

(ii) $H_K^q(X, \mathcal{F}\otimes_{\mathcal{O}} \Omega^p)$ with its natural topology is the strong dual of

$H^0(K, \mathcal{E}xt_{\mathcal{O}}^{n-q}(\mathcal{F}, \mathcal{O})\otimes_{\mathcal{O}} \Omega^{n-p})$.

Proof. First we observe that $\Gamma(K, \mathcal{F})$ is always Hausdorff. For, if

$$\mathcal{O}^q \to \mathcal{O}^p \to \mathcal{F} \to 0$$

is a resolution on K, then, since every submodule of $\mathcal{O}_x^p$ is closed in the topology of simple convergence of power series coefficients

[12, p. 152, Th. 6.3.5], the image of

$$\Gamma(K, \mathcal{O}^q) \to \Gamma(K, \mathcal{O}^p)$$

is closed.

Let $\mathrm{codh}_K \mathcal{F} = \inf_{x \in K} \mathrm{codh}_x \mathcal{F}$. We are going to prove the theorem by descending induction on $\mathrm{codh}_K \mathcal{F}$. If $\mathrm{codh}_K \mathcal{F} \geqq n$, then $\mathcal{F}$ is locally free on some Stein open neighborhood of K and the theorem follows from Theorem (5.9).

Assume $\mathrm{codh}_K \mathcal{F} < n$. By Cartan's theorem A we get an exact sequence

$$(*) \qquad 0 \to \mathcal{G} \to \mathcal{O}^s \to \mathcal{F} \to 0$$

on some Stein open neighborhood of K (which without loss of generality we can assume to be X). The sequence

$$(**) \qquad 0 \leftarrow \mathcal{E}xt^1_{\mathcal{O}}(\mathcal{F}, \mathcal{O}) \leftarrow \mathcal{G}^* \leftarrow \mathcal{O}^s \leftarrow \mathcal{F}^* \leftarrow 0$$

is exact. By tensoring (*) and (**) respectively with Ω^p and Ω^{n-p} and using Theorem (5.9) and Cartan's theorem B, we get the exact sequences of continuous maps

$$0 \to H^{n-1}_K(X, \mathcal{F} \otimes_{\mathcal{O}} \Omega^p) \to H^n_K(X, \mathcal{G} \otimes_{\mathcal{O}} \Omega^p) \xrightarrow{\alpha} H^n_K(X, \mathcal{O}^s \otimes_{\mathcal{O}} \Omega^p) \to H^n_K(X, \mathcal{F} \otimes_{\mathcal{O}} \Omega^p) \to 0$$

$$0 \leftarrow H^0(K, \mathcal{E}xt^1_{\mathcal{O}}(\mathcal{F}, \mathcal{O}) \otimes_{\mathcal{O}} \Omega^{n-p}) \leftarrow H^0(K, \mathcal{G}^* \otimes_{\mathcal{O}} \Omega^{n-p}) \xleftarrow{\alpha^t}$$

$$H^0(K, \mathcal{O}^s \otimes_{\mathcal{O}} \Omega^{n-p}) \leftarrow H^0(K, \mathcal{F}^* \otimes_{\mathcal{O}} \Omega^{n-p}) \leftarrow 0.$$

Since

$$H^0(K, \mathcal{E}xt^1_{\mathcal{O}}(\mathcal{F}, \mathcal{O}) \otimes_{\mathcal{O}} \Omega^{n-p})$$

is Hausdorff, α^t and α have closed range by (5.6).

By induction hypothesis $H^0(K, \mathcal{G}^* \otimes \Omega^{n-p})$ and $H^0(K, \mathcal{O}^s \otimes_{\mathcal{O}} \Omega^{n-p})$ are respectively the strong duals of the (FS)-spaces $H^n_K(X, \mathcal{G} \otimes_{\mathcal{O}} \Omega^p)$ and $H^n_K(X, \mathcal{O}^s \otimes_{\mathcal{O}} \Omega^p)$. Since $H^0(K, \mathcal{F}^* \otimes_{\mathcal{O}} \Omega^p)$ is Hausdorff, by Proposition (5.7) and the generalized closed graph theorem [13, p. 304, Prop. 10] $H^{n-1}_K(X, \mathcal{F} \otimes_{\mathcal{O}} \Omega^p)$ and $H^n_K(X, \mathcal{F} \otimes_{\mathcal{O}} \Omega^p)$ are respectively the strong duals of the (DFS)-spaces $H^0(K, \mathcal{E}xt^1_{\mathcal{O}}(\mathcal{F}, \mathcal{O}) \otimes_{\mathcal{O}} \Omega^{n-p})$ and $H^0(K, \mathcal{F}^* \otimes_{\mathcal{O}} \Omega^{n-p})$. Finally the duality between the (FS)-space $H^q_K(X, \mathcal{F} \otimes_{\mathcal{O}} \Omega^p)$ and the (DFS)-space $H^0(K, \mathcal{E}xt^{n-q}_{\mathcal{O}}(\mathcal{F}, \mathcal{O}) \otimes_{\mathcal{O}} \Omega^{n-p})$ for $q \leqq n - 2$ follows from the induction hypothesis, the topological isomorphism

$$H^q_K(X, \mathcal{F} \otimes_{\mathcal{O}} \Omega^p) \approx H^{q+1}_K(X, \mathcal{G} \otimes_{\mathcal{O}} \Omega^p)$$

and the sheaf-isomorphism

$$\mathcal{E}xt^{n-q}_{\mathcal{O}}(\mathcal{F}, \mathcal{O}) \approx \mathcal{E}xt^{n-(q+1)}_{\mathcal{O}}(\mathcal{G}, \mathcal{O}),$$

where the sheaf-isomorphism is a consequence of (*) and the topological isomorphism is a consequence of (*) and [13, p. 301, Th. 4].

Q.E.D.

(5.13) <u>Corollary</u>. Let X be a Stein manifold of dimension n and $\mathcal{F}$ a coherent analytic sheaf on X. Then $H^q_c(X, \mathcal{F} \otimes_{\mathcal{O}} \Omega^p)$ is the topological dual of the (FS)-space $H^0(X, \mathcal{E}xt^{n-q}_{\mathcal{O}}(\mathcal{F}, \mathcal{O}) \otimes_{\mathcal{O}} \Omega^{n-p})$.

Proof. Follows from Theorem (5.12) by the fact that

$$H^q_c(X, \mathcal{F}\otimes_{\mathcal{O}} \Omega^p) = \varinjlim\{H^q_K(X, \mathcal{F}\otimes_{\mathcal{O}} \Omega^p) \mid K \text{ is a Stein compact subset of } X\}.$$

Q.E.D.

§6 ρ-convexity

In the rest of these lecture notes we will use the following notations. If $\alpha_1,\dots,\alpha_n$ are positive numbers, then

$$K^n(\alpha_1,\dots,\alpha_n) = \{(z_1,\dots,z_n) \in \mathbb{C}^n \,\big|\, |z_i| < \alpha_i \text{ for } 1 \leqq i \leqq n\}.$$

If $0 \leqq \alpha_i' < \alpha_i$ for $1 \leqq i \leqq n$, then

$$G^n(\alpha_1',\dots,\alpha_n';\alpha_1,\dots,\alpha_n) =$$

$$\{(z_1,\dots,z_n) \in K^n(\alpha_1,\dots,\alpha_n) \,\big|\, |z_i| > \alpha_i' \text{ for some } 1 \leqq i \leqq n\}.$$

${}_n\mathcal{O}$ = the sheaf of germs of holomorphic functions on $\mathbb{C}^n$.

(6.1) Suppose φ is a C^2 real-valued function on an open subset G of $\mathbb{C}^n$ and $z^0 = (z_1^0,\dots,z_n^0) \in G$. We associate to φ and z^0 the following polynomial

$$f(z) = \varphi(z^0) + 2\Sigma_{i=1}^n (z_i - z_i^0)\frac{\partial\varphi}{\partial z_i}(z^0) + \Sigma_{i,j=1}^n (z_i - z_i^0)(z_j - z_j^0)\frac{\partial^2\varphi}{\partial z_i \partial z_j}(z^0).$$

We call f the associated quadratic polynomial of φ at z^0. When the coordinates system undergoes an affine transformation, the associated quadratic polynomial remains unchanged.

A C^2 real-valued function φ is said to be strongly ρ-convex on G if at every point of G the hermitian matrix

$$L = \left(\frac{\partial^2\varphi}{\partial z_i \partial \bar{z}_j}\right)_{i,j=1,\dots,n}$$

has at least $n-\rho+1$ positive eigenvalues. This definition is independent of the coordinates system. If we choose another coordinates system $\zeta_1,\ldots,\zeta_n$ on G, then at every point of G the hermitian matrix

$$\left(\frac{\partial^2\varphi}{\partial\zeta_i\partial\bar{\zeta}_j}\right)_{i,j=1,\ldots,n}$$

has the same number of positive eigenvalues as L.

A strongly ρ-convex function φ is said to be <u>regular</u> in $z_1,\ldots,z_n$ if at every point of G the hermitian matrix

$$\left(\frac{\partial^2\varphi}{\partial z_i\partial\bar{z}_j}\right)_{i,j=\rho,\ldots,n}$$

is positive definite. If φ is regular in $z_1,\ldots,z_n$, then for every relatively compact open subset G' of G there exists $\varepsilon > 0$ such that $\varphi|G'$ is regular in $z_1',\ldots,z_n'$ if

$$z_i' = \Sigma_{j=1}^n a_{ij}z_j$$

and

$$|a_{ij}| < \varepsilon \text{ for } i \neq j$$

$$|a_{ii} - 1| < \varepsilon.$$

A function ψ on a complex space X is said to be strongly <u>ρ-convex</u> if for every point x of X there exist

(i) an open neighborhood U of x in X,

(ii) a biholomorphic map σ from U onto a complex subspace of an open subset G of some $\mathbb{C}^n$,

(iii) a strongly ρ-convex function φ on G such that $\psi|U = \varphi \circ \sigma$.

An open subset D of a complex space X is said to be <u>strongly ρ-concave</u> at a point x of X if there exist

(i) an open neighborhood U of x in X,

(ii) a strongly ρ-convex function ψ on U such that $\psi(x) = 0$ and $D \cap U = \{y \in U | \psi(y) > 0\}$.

We will be mainly interested in the case where x is a boundary point of D.

(6.2) <u>Lemma</u>. Suppose φ is a strongly 1-convex function on an open neighborhood G of z^0 in $\mathbb{C}^n$. Let f be the associated quadratic polynomial of φ at z^0. Then there exist $\gamma > 0$ and an open neighborhood U of z^0 in G such that

$$\varphi(z) \geqq \operatorname{Re} f(z) + \gamma \Sigma_{i=1}^n |z_i - z_i^0|^2$$

for $z \in U$.

Proof. In an open neighborhood G' of z^0 in G we have the following Taylor series expansion:

$$\varphi(z) = \operatorname{Re} f(z) + \Sigma_{i,j=1}^n (z_i - z_i^0)\overline{(z_j - z_j^0)} \frac{\partial^2 \varphi}{\partial z_i \partial \overline{z}_j}(z^0) + \varepsilon(z),$$

where

$$\lim_{z \to z^0} \frac{\varepsilon(z)}{\Sigma_{i=1}^n |z_i - z_i^0|^2} = 0.$$

Since

$$\left(\frac{\partial^2 \varphi}{\partial z_i \partial \bar{z}_j}(z^0)\right)_{i,j=1,\ldots,n}$$

is positive definite,

$$\Sigma_{i,j=1}^n (z_i - z_i^0)\overline{(z_j - z_j^0)} \frac{\partial^2 \varphi}{\partial z_i \partial \bar{z}_j}(z^0) \geqq 2\gamma\, \Sigma_{i=1}^n |z_i - z_i^0|^2$$

for some $\gamma > 0$. There exists an open neighborhood U of z^0 in G' such that

$$\frac{\varepsilon(z)}{\Sigma_{i=1}^n |z_i - z_i^0|^2} \leqq \gamma$$

for $z \in U - \{z^0\}$. Q.E.D.

(6.3) Lemma. Suppose φ is a strongly 1-convex function on a complex space X whose branches are all positive-dimensional. Then φ cannot achieve a maximum on X.

Proof. We can assume that X is reduced and irreducible. Suppose φ achieves its maximum on X. By (0.16) φ is constant on X.

Take a regular point x of X and take a coordinates system

$\zeta_1,\ldots,\zeta_k$ on an open neighborhood U of x in X, where k = dim X. Since φ is strongly 1-convex, it is easily seen that the hermitian matrix

$$L = \left(\frac{\partial^2\varphi}{\partial\zeta_i\partial\overline{\zeta}_j}\right)_{i,j=1,\ldots,k}$$

is positive definite at every point of U. However, since φ is constant, L must be identically zero on U. Contradiction. Q.E.D.

(6.4) Lemma. Suppose $G \subset\subset \tilde{G}$ are open subsets of $\mathbb{C}^n$ and φ is a strongly 1-convex function on $\tilde{G}$. Suppose V is a subvariety in $\tilde{G} \cap \{\varphi > c\}$ for some $c \in \mathbb{R}$ and $V \subset G^-$. Then dim V $\leqq$ 0.

Proof. Suppose V_1 is a positive-dimensional branch of V. Take $x_0 \in V_1$. Let M be the supremum of φ on V_1. M is achieved by φ at some point of V_1, because M is equal to the supremum of φ on the compact set $G^- \cap \{\varphi \geqq \varphi(x_0)\} \cap V_1$. This contradicts Lemma (6.3). Q.E.D.

(6.5) Lemma. Suppose $\alpha_1 > 0,\ldots,\alpha_\rho > 0$, $\alpha_{\rho+1} > \alpha'_{\rho+1} > 0,\ldots,\alpha_n > \alpha'_n > 0$ and φ is strongly $(\rho+1)$-convex on $K^n(\alpha_1,\ldots,\alpha_n)$ such that φ is regular in $z_{\rho+1},\ldots,z_n$. Let $D = \{\varphi > 0\}$.

(a) If Y is a subvariety in D and

$$Y \cap K^\rho(\alpha_1,\ldots,\alpha_\rho) \times G^{n-\rho}(\alpha'_{\rho+1},\ldots,\alpha'_n;\alpha_{\rho+1},\ldots,\alpha_n) = \emptyset ,$$

then dim Y $\leqq \rho$.

(b) Suppose f is a holomorphic function on $K^n(\alpha_1,\dots,\alpha_n)$ and $\beta > 0$. If Y is a subvariety in $D \cap \{|f| < \beta\}$ and

$$Y \cap K^\rho(\alpha_1,\dots,\alpha_\rho) \times G^{n-\rho}(\alpha'_{\rho+1},\dots,\alpha'_n;\alpha_{\rho+1},\dots,\alpha_n) = \emptyset ,$$

then $\dim Y \leqq \rho + 1$.

Proof. (a) Take arbitrarily $y \in Y$. Let $Z = Y \cap \{z_1=z_1(y),\dots,z_\rho = z_\rho(y)\}$. Identify $\{z_1 = z_1(y),\dots,z_\rho = z_\rho(y)\}$ with $\mathbb{C}^{n-\rho}$ and apply Lemma (6.4) with $G = K^{n-\rho}(\alpha'_{\rho+1},\dots,\alpha'_n)$ and $\tilde{G} = K^{n-\rho}(\alpha_{\rho+1},\dots,\alpha_n)$. We conclude that $\dim Z \leqq 0$. Since y is arbitrary, by (0.17) $\dim Y \leqq \rho$.

(b) Take arbitrarily $y \in Y$. Let $Z = Y \cap \{f=f(y)\}$. Then Z is a subvariety in D and is disjoint from $K^\rho(\alpha_1,\dots,\alpha_\rho) \times G^{n-\rho}(\alpha'_{\rho+1},\dots,\alpha'_n;$ $\alpha_{\rho+1},\dots,\alpha_n)$. By (a), $\dim Z \leqq \rho$. Since y is arbitrary, by (0.17) $\dim Y \leqq \rho + 1$. Q.E.D.

(6.6) Proposition. Suppose φ is a strongly ρ-convex function on a complex space X and every branch of X has dimension $\geqq \rho$. Then φ cannot achieve its maximum on X.

Proof. Suppose φ achieves its maximum at some point x_0 of X. By replacing X by an open neighborhood of x_0, we can assume w.l.o.g. that X is a complex subspace of an open subset G of some $\mathbb{C}^n$ and that φ is the restriction to X of a strongly ρ-convex function ψ on G. We can also assume that ψ is regular in $z_\rho, \ldots, z_n$. Let $E = \{z_1 = z_1(x^0), \ldots, z_{\rho-1} = z_{\rho-1}(x^0)\}$. By (0.17) every branch of $E \cap X$ is positive-dimensional, because every branch of X has dimension $\geqq \rho$. $\varphi|E \cap X$ is strongly 1-convex on $E \cap X$ and achieves its maximum at x^0, contradicting Lemma (6.3). Q.E.D.

(6.7) <u>Proposition</u>. Suppose D is an open subset of a complex space X and D is strongly ρ-concave at a point x of X. Suppose V is a subvariety defined in an open neighborhood U of x in X such that $x \in V$ and $V \cap D = \emptyset$. Then $\dim_x V \leqq \rho - 1$.

Proof. Since the problem is local in nature, we can assume that $D = \{\varphi > 0\}$ and $\varphi(x) = 0$ for some strongly ρ-convex function φ on X. Suppose V_1 is a branch of V which passes through x and has dimension $\geqq \rho$. Since $\varphi(x) = 0$ and $\varphi \leqq 0$ on V_1, $\varphi|V_1$ achieves its maximum at x, contradicting Proposition (6.6). Q.E.D.

(6.8) <u>Proposition</u>. Suppose D is an open subset of a complex space X and D is strongly ρ-concave at a point x of X. Suppose V is a subvariety in D and every branch of V has dimension $\geqq \rho$. If U_i is an open neighborhood of x in X and $\tilde{V}_i$ is a subvariety of $D \cup U_i$ such that $\tilde{V}_i$ extends V and every branch of $\tilde{V}_i$ has dimension $\geqq \rho$ $(i=1,2)$, then the germ of $\tilde{V}_1$ at x agrees with the germ of $\tilde{V}_2$ at x.

Proof. Take a Stein open neighborhood G of x in $U_1 \cap U_2$. Let F be the set of all holomorphic functions on G which vanish identically on $V \cap G$. Let V^* be the subvariety of G defined by the vanishing of all members of F. Then V* equals the union of branches of $\tilde{V}_i \cap G$ which intersect D (i=1,2). Hence by Proposition (6.7) the germ of V* at x agrees with the germ of $\tilde{V}_i$ at x (i=1,2). The germ of $\tilde{V}_1$ at x is therefore the same as the germ of $\tilde{V}_2$ at x. Q.E.D.

§7 Extension of analytic covers

All complex spaces in this paragraph are reduced.

(7.1) Suppose U is an open subset of $\mathbb{C}^m$, X is a complex space, and π: $X \to U$ is a λ-sheeted analytic cover with critical set A (0.15). Suppose f is a holomorphic function on X.

For $z' \in U - A$ let $\pi^{-1}(z') = \{z^{(1)},\ldots,z^{(\lambda)}\}$. The ith elementary symmetric polynomial $\alpha_i(z')$ of $f(z^{(1)}),\ldots,f(z^{(\lambda)})$ depends only on z' and is a holomorphic function on U - A. Since π is proper, $\alpha_i(z')$ is locally bounded on U and hence can be extended uniquely to a holomorphic function $\tilde{\alpha}_i$ on U. Let $a_i = (-1)^{\lambda-i}\tilde{\alpha}_{\lambda-i}$.

Introduce the polynomial

$$P_f(z;Z) = Z^\lambda + \Sigma_{i=0}^{\lambda-1} a_i(z)Z^i.$$

$P_f(z;Z)$ has the property that $P_f(\pi(x);\ f(x)) \equiv 0$ for $x \in X$. In fact, $P_f(z;Z)$ is the only monic polynomial of degree λ with coefficients in $\Gamma(U,{}_m\mathcal{O})$ which enjoys this property.

Denote by $P_f'(z;Z)$ the derivative of $P_f(z;Z)$ with respect to Z. We have

$$\frac{P_f(z;Z) - P_f(z;Z')}{Z - Z'} = \Sigma_{i,j=0}^{\lambda-1} b_{ij}(z)\ Z^i(Z')^j\ ,$$

where $b_{ij} \in \Gamma(U,{}_m\mathcal{O})$. By substituting $z = z'$, $Z = f(z^{(k)})$, and $Z' = f(z^{(\ell)})$, we obtain

$$\begin{cases} 0 = \Sigma_{i,j=0}^{\lambda-1} b_{ij}(z')\ f(z^{(k)})^i\ f(z^{(\ell)})^j \text{ for } \ell \neq k \\ \\ P_f'(z';f(z^{(k)})) = \Sigma_{i,j=0}^{\lambda-1} b_{ij}(z')\ f(z^{(k)})^{i+j}\ . \end{cases}$$

Suppose g is another holomorphic function on X. By multiplying the preceding two equations by $g(z^{(\ell)})$ and $g(z^{(k)})$ respectively and adding up all the λ equations (when ℓ runs through $\{1,\ldots,\lambda\}-\{k\}$), we derive

$$(*) \qquad g(z^{(k)})P_f'(z';f(z^{(k)})) = \Sigma_{i,j=0}^{\lambda-1} b_{ij}(z')\ f(z^{(k)})^i\ \Sigma_{\ell=1}^{\lambda} f(z^{(\ell)})^j g(z^{(\ell)}).$$

The function $\Sigma_{j=0}^{\lambda-1} b_{ij}(z')\ \Sigma_{\ell=1}^{\lambda} f(z^{(\ell)})^j\ g(z^{(\ell)})$ depends only on z' and is holomorphic on U - A. Since it is locally bounded on U, it can be extended uniquely to a holomorphic function c_i on U.

Introduce the polynomial

$$T_{f,g}(z;Z) = \Sigma_{i=0}^{\lambda-1} c_i(z)Z^i.$$

The following identity

$$(\dagger) \qquad g(x)P_f'(\pi(x);f(x)) \equiv T_{f,g}(\pi(x);f(x))$$

holds for $x \in X$. For, with the substitution $z^{(k)} = x$ and $z' = \pi(x)$, (*) implies (†) for $x \in X - \pi^{-1}(A)$, and by continuity it follows that (†) holds for all $x \in X$.

(7.2) Suppose X is the same as in (7.1). We assume in addition the following:

(i) X is a subvariety of $U \times \mathbb{C}^n$,

(ii) $\pi\colon X \to U$ is induced by the natural projection $\Pi\colon \mathbb{C}^m \times \mathbb{C}^n \to \mathbb{C}^m$,

(iii) U is connected.

Let $w_1, \ldots, w_n$ be the coordinates of $\mathbb{C}^n$. We can choose $z^* \in U - A$ and $\mu_1, \ldots, \mu_n \in \mathbb{C}$ such that the function $\mu_1 w_1 + \ldots + \mu_n w_n$ separates all the λ points in $\pi^{-1}(z^*)$.

Denote by $P_0(z;Z)$ the polynomial $P_f(z;Z)$ when $f = (\mu_1 w_1 + \ldots + \mu_n w_n)|X$. Denote by $P_k(z;Z)$ the polynomial $P_f(z;Z)$ when $f = w_k|X$. Denote by $T_k(z;Z)$ the polynomial $T_{f,g}(z;Z)$ when $f = (\mu_1 w_1 + \ldots + \mu_n w_n)|X$ and $g = w_k|X$.

(7.2.1) <u>Lemma</u>. X is equal to the m-dimensional component of the subvariety X' of $U \times \mathbb{C}^n$ defined by the following equations (where $z \in U$ and $(w_1, \ldots, w_n) \in \mathbb{C}^n$):

$$\begin{cases} P_0(z;\mu_1 w_1 + \ldots + \mu_n w_n) = 0 \\ P_k(z;w_k) = 0 \qquad (1 \leqq k \leqq n) \\ w_\ell P_0'(z;\mu_1 w_1 + \ldots + \mu_n w_n) = T_\ell(z;\mu_1 w_1 + \ldots + \mu_n w_n) \quad (1 \leqq \ell \leqq n). \end{cases}$$

Proof. By (*) and (†) the above equations are satisfied when $(z;w_1, \ldots, w_n) \in X$. Since $\dim X = m$, X is contained in the m-dimensional component of X'.

Let B be the subvariety of U defined by the vanishing of the discriminant of the polynomial $P_0(z;Z)$. It is clear that $A \subset B$. Since $\mu_1 w_1 + \ldots + \mu_n w_n$ separates all the λ points in $\pi^{-1}(z^*)$, $z^* \notin B$. B is of codimension $\geqq 1$ in U.

Take arbitrarily $(z';w_1',\ldots,w_n') \in X'$ with $z' \in U - B$. Let $\pi^{-1}(z') = \{z^{(1)},\ldots,z^{(\lambda)}\}$ and $w_j^{(i)} = w_j(z^{(i)})$. Since $\mu_1 w_1^{(i)} + \ldots + \mu_n w_n^{(i)}$ $(1 \leqq i \leqq \lambda)$ are λ distinct roots of the polynomial $P_0(z;Z)$ of degree λ, $\mu_1 w_1' + \ldots + \mu_n w_n' = \mu_1 w_1^{(i)} + \ldots + \mu_n w_n^{(i)}$ for some i. From

$$w_\ell' \, P_0'(z';\mu_1 w_1' + \ldots + \mu_n w_n') = T_\ell(z';\mu_1 w_1' + \ldots + \mu_n w_n')$$

and

$$w_\ell^{(i)} P_0'(z';\mu_1 w_1^{(i)} + \ldots + \mu_n w_1^{(i)}) = T_\ell(z';\mu_1 w_1^{(i)} + \ldots + \mu_n w_n^{(i)})$$

it follows that $w_\ell' = w_\ell^{(i)}$ for $1 \leqq \ell \leqq n$. Hence $(z';w_1',\ldots,w_n') = z^{(i)} \in X$. We conclude that

$$X \cap (U-B) \times \mathbb{C}^n = X' \cap (U-B) \times \mathbb{C}^n.$$

By virtue of the equations $P_k(z;w_k) = 0$ $(1 \leqq k \leqq n)$, $X' \cap B \times \mathbb{C}^n$ has dimension $\leqq m - 1$. Therefore X agrees with the m-dimensional component of X'. Q.E.D.

(7.2.2) Lemma. Suppose $\tilde{U}$ is a connected open subset of $\mathbb{C}^m$ containing U such that the restriction map $\Gamma(\tilde{U}, {}_m\mathcal{O}) \to \Gamma(U, {}_m\mathcal{O})$ is bijective. Then X can be extended uniquely to a subvariety $\tilde{X}$ of $\tilde{U} \times \mathbb{C}^n$ which is an analytic cover over $\tilde{U}$ under the projection $\tilde{\pi}: \tilde{X} \to \tilde{U}$ induced by Π.

Proof. The coefficients of $P_k(z;Z)$ $(0 \leq k \leq n)$ and $T_\ell(z;Z)$ $(1 \leq \ell \leq n)$ can be uniquely extended to holomorphic functions on $\tilde{U}$. Using these new holomorphic functions as coefficients, we form polynomials $\tilde{P}_k(z;Z)$ and $\tilde{T}_\ell(z;Z)$. Let $\tilde{X}$ be the m-dimensional component of the subvariety of $\tilde{U} \times \mathbb{C}^m$ defined by the following equations:

$$\begin{cases} \tilde{P}_0(z;\mu_1 w_1+\ldots+\mu_n w_n) = 0 \\ \tilde{P}_k(z;w_k) = 0 \quad (1 \leq k \leq n) \\ w_\ell \tilde{P}_0'(z;\mu_1 w_1+\ldots+\mu_n w_n) = \tilde{T}_\ell(z;\mu_1 w_1+,,,+\mu_n w_n) \quad (1 \leq \ell \leq n). \end{cases}$$

By Lemma (7.2.1), $\tilde{X}$ is the required extension of X. Q.E.D.

(7.2.3) Lemma. Suppose $\tilde{f}$ is a holomorphic function on $\tilde{X}$ and $f = \tilde{f}|X$. Then $\|\tilde{f}\|_{\tilde{X}} = \|f\|_X$.

Proof. Obviously $\|f\|_X \leq \|\tilde{f}\|_{\tilde{X}}$. Suppose $|\tilde{f}(x)| > \|f\|_X$ for some $x \in \tilde{X}$. Let $\tilde{\pi}^{-1}(\tilde{\pi}(x)) = (x^{(1)},\ldots,x^{(\nu)})$. We can assume (after renumbering) that

$$|\tilde{f}(x^{(1)})| \leq \ldots \leq |\tilde{f}(x^{(k)})| < |\tilde{f}(x^{(k+1)})| = \ldots = |\tilde{f}(x^{(\nu)})| ,$$

where k may be 0. Choose A such that

$$\|f\|_X < A \text{ and } |\tilde{f}(x^{(k)})| < A < |\tilde{f}(x^{(k+1)})|.$$

Consider the coefficient $a_1^{(\ell)}(z)$ of Z in the polynomial $P_{(\frac{\tilde{f}}{A})^\ell}(z;Z)$.

When $\ell \to \infty$, $a_1^{(\ell)}(z) \to 0$ uniformly on U, but $|a_1^{(\ell)}(\pi(x))| \to \infty$, contradicting that the restriction map $\Gamma(\tilde{U}, {}_m\mathcal{O}) \to \Gamma(U, {}_m\mathcal{O})$ is an isomorphism of Frechet spaces. Q.E.D.

Since every branch of $\tilde{X}$ intersects X, by (0.16) we have the following two corollaries:

(7.2.4) <u>Corollary</u>. If $|f(x)| < M$ for every $x \in X$, then $|\tilde{f}(x)| < M$ for every $x \in \tilde{X}$.

(7.2.5) <u>Corollary</u>. Suppose $f_1, \ldots, f_\ell$ are holomorphic functions on $\mathbb{C}^n$ and $P = \{w \in \mathbb{C}^n \mid |f_i(w)| < \alpha_i \text{ for } 1 \leqq i \leqq \ell\}$, where $\alpha_i > 0$. If $X \subset U \times P$, then $\tilde{X} \subset \tilde{U} \times P$.

(7.2.6) <u>Proposition</u>. Suppose P is the same as in Corollary (7.2.5). Let $\gamma: P \to \mathbb{C}^m$ be a holomorphic map. Let $Q = P \cap \gamma^{-1}(U)$ and $\tilde{Q} = P \cap \gamma^{-1}(\tilde{U})$. Suppose Y is a subvariety in Q such that $\gamma|Y$ makes Y an analytic cover over U. Then Y can be extended uniquely to a subvariety $\tilde{Y}$ in $\tilde{Q}$ such that $\gamma|\tilde{Y}$ makes $\tilde{Y}$ an analytic cover over $\tilde{U}$.

Proof. Let $\Phi: P \to \mathbb{C}^m \times \mathbb{C}^n$ be defined by $\Phi(x) = (\gamma(x), x)$. $\Phi(Y)$ is a subvariety of $U \times P$, because, if Y is the set of common zeros of holomorphic functions g_i $(i \in I)$ on Q, then

$$\Phi(Y) = \{(z,w) \in \mathbb{C}^m \times \mathbb{C}^n | z - \gamma(w) = 0,\ g_i(w) = 0 \text{ for } i \in I\}.$$

Since $\gamma | Y$ makes Y an analytic cover over U and $\Pi \circ \Phi = \gamma$, $\Pi | \Phi(Y)$ makes $\Phi(Y)$ an analytic cover over U. Hence $\Phi(Y)$ is a subvariety in $U \times \mathbb{C}^n$. By Lemma (7.2.2), $\Phi(Y)$ can be extended to a subvariety Y^* in $\tilde{U} \times \mathbb{C}^n$ such that $\Pi | Y^*$ makes Y^* an analytic cover over $\tilde{U}$. By Corollary (7.2.5), $Y^* \subset \tilde{U} \times P$. $\tilde{Y} = \Phi^{-1}(Y^*)$ is the desired extension for Y. Q.E.D.

§8 Subvariety extension. Projection Lemma

Suppose G is an open neighborhood of 0 in $\mathbb{C}^n$ and φ is a strongly ρ-convex function on G with $\varphi(0) = 0$. Let $D = \{\varphi > 0\}$. Suppose X is a subvariety of pure dimension $\rho + k$ in D, where $k \geqq 1$. Let f be the associated quadratic polynomial of φ at 0.

(8.1) Suppose f does not vanish identically on any branch of X.

(8.1.1) Lemma. After a homogeneous affine transformation of the coordinates system of $\mathbb{C}^n$ there exist

$$\alpha_1 > 0,\ldots,\alpha_{\rho-1} > 0,\ \alpha_\rho > \alpha'_\rho > 0,\ldots,\alpha_n > \alpha'_n > 0,\ \beta > 0$$

such that

(i) $K^n(\alpha_1,\ldots,\alpha_n) \subset G$,

(ii) $K^{\rho+k-1}(\alpha_1,\ldots,\alpha_{\rho+k-1}) \times G^{n-\rho-k+1}(\alpha'_{\rho+k},\ldots,\alpha'_n;\alpha_{\rho+k},\ldots,\alpha_n) \cap$

$\{|f| < \beta\}$ is contained in $D - X$,

(iii) $K^{\rho-1}(\alpha_1,\ldots,\alpha_{\rho-1}) \times G^{n-\rho+1}(\alpha'_\rho,\ldots,\alpha'_n;\alpha_\rho,\ldots,\alpha_n) \cap \{|f| < \beta\}$

is contained in D,

(iv) $K^n(\alpha_1,\ldots,\alpha_n) \cap \{|f - \frac{\beta}{2}| < \frac{\beta}{4}\} \subset D$, and

(v) φ is regular in $z_1,\ldots,z_n$ on $K^n(\alpha_1,\ldots,\alpha_n)$.

Proof. If $0 \notin X$, everything is trivial. So we assume $0 \in X$. Since

$$\dim_0 X \cap \{f = 0\} = \rho + k - 1,$$

according to (0.18), after a homogeneous affine transformation of the coordinates system of $\mathbb{C}^n$ and after a shrinking of G, we can assume that

(i) φ is regular in $z_1, \ldots, z_n$, and

(ii) $\dim X \cap \{f = z_1 = \ldots = z_{\rho+k-1} = 0\} = 0$.

By (0.19), after a further shrinking of G, we can assume that the map $X \to \mathbb{C}^{\rho+k}$ defined by $(z_1, \ldots, z_{\rho+k-1}, f)$ has discrete fibers.

Let $E = \{z_1 = \ldots = z_{\rho-1} = 0\}$. Since $\varphi | E \cap G$ is strongly 1-convex and $f|E$ is the associated quadratic polynomial of $\varphi | E \cap G$ at 0, by Lemma (6.2) we can choose $\gamma > 0$ and an open neighborhood U of 0 in $E \cap G$ such that

(*) $$\varphi(z) \geqq \operatorname{Re} f(z) + \gamma \, \Sigma_{i=\rho}^{n} |z_i|^2$$

for $z \in U$.

Since $X \cap \{z_1 = \ldots = z_{\rho+k-1} = f = 0\}$ is discrete, we can choose

$$\alpha_{\rho+k} > \alpha'_{\rho+k} > 0, \ldots, \alpha_n > \alpha'_n > 0$$

such that

(i) $\{z_1 = \ldots = z_{\rho+k-1} = 0\} \cap \{|z_{\rho+k}| \leqq \alpha_{\rho+k}, \ldots, |z_n| \leqq \alpha_n\}$ is contained in U, and

(ii) $A: = \bigcup_{i=\rho+k}^{n}\{z_1=\ldots=z_{\rho+k-1}=f = 0,\ \alpha_i' \leqq |z_i| \leqq \alpha_i,\ |z_j| \leqq \alpha_j,$

$\rho + k \leqq j \leqq n\}$ is disjoint from X.

By virtue of (*), $A \subset D$. Since X is a closed subset of D, D - X is an open neighborhood of A. We can find positive numbers

$$\alpha_1^*,\ldots,\alpha_{\rho-1}^*,\alpha_\rho,\ldots,\alpha_{\rho+k-1},\beta$$

such that

(i) $K^n(\alpha_1^*,\ldots,\alpha_{\rho-1}^*,\alpha_\rho,\ldots,\alpha_n) \subset G,$

(ii) $\beta < \frac{\gamma}{4}\inf\{\alpha_i^2 | \rho \leqq i \leqq n\}$

(iii) $\{z_1=\ldots=z_{\rho-1}=0,\ |z_j| \leqq \alpha_j \text{ for } \rho \leqq j \leqq n\} \subset U$, and

(iv) $\bigcup_{i=\rho+k}^{n} \{|z_1| < \alpha_1^*,\ldots,|z_{\rho-1}| < \alpha_{\rho-1}^*,\ |z_\rho| < \alpha_\rho,\ldots,|z_{\rho+k-1}| <$

$\alpha_{\rho+k-1},\ |f| < \beta,\ \alpha_i' \leqq |z_i| \leqq \alpha_i,\ |z_j| \leqq \alpha_j \text{ for } \rho + k \leqq j \leqq n\}$ is contained in D - X.

Let $\alpha_i' = \frac{\alpha_i}{2}$ for $\rho \leqq i \leqq \rho + k - 1$. Let C be the union of the following two sets:

$$\bigcup_{i=\rho}^{n} \{z_1=\ldots=z_{\rho-1}=0,\ |f| \leqq \beta,\ \alpha_i' \leqq |z_i| \leqq \alpha_i,\ |z_j| \leqq \alpha_j \text{ for } \rho\leqq j\leqq n\},$$

$$\{z_1=\ldots=z_{\rho-1}=0,\ |f - \tfrac{\beta}{2}| \leqq \tfrac{\beta}{4},\ |z_\rho| \leqq \alpha_\rho,\ldots,|z_n| \leqq \alpha_n\}.$$

Since $\beta < \frac{\gamma}{4}\inf\{\alpha_i^2 | \rho \leqq i \leqq n\}$, by (*) $C \subset D$. Since C is compact, we can choose

$$0 < \alpha_1 < \alpha_1^*,\ldots,0 < \alpha_{\rho-1} < \alpha_{\rho-1}^*$$

such that conditions (iii) and (iv) of the statement of the Lemma are satisfied. Q.E.D.

(8.1.2) <u>Corollary</u>. Let $Q = K^n(\alpha_1,\ldots,\alpha_n) \cap \{|f| < \beta\}$ and $\sigma\colon Q \to \mathbb{C}^{\rho+k}$ be defined by $z_1,\ldots,z_{\rho+k-1},f$. Let $B = \{z \in \mathbb{C} \mid |z - \frac{\beta}{2}| < \frac{\beta}{4}\}$ and H be the union of

$$K^{\rho-1}(\alpha_1,\ldots,\alpha_{\rho-1}) \times G^k(\alpha'_\rho,\ldots,\alpha'_{\rho+k-1}; \alpha_\rho,\ldots,\alpha_{\rho+k-1}) \times K^1(\beta)$$

and

$$K^{\rho+k-1}(\alpha_1,\ldots,\alpha_{\rho+k-1}) \times B.$$

Then

(a) σ maps $\sigma^{-1}(H) \cap X$ properly to H, and

(b) every branch of $Q \cap X$ intersects $\sigma^{-1}(H) \cap X$.

Hence $Q \cap X$ has only a finite number of branches.

Proof. Let

$$W = K^{\rho+k-1}(\alpha_1,\ldots,\alpha_{\rho+k-1}) \times \overline{K}^{n-\rho-k+1}(\alpha'_{\rho+k},\ldots,\alpha'_n) \cap \{|f| < \beta\}.$$

Clearly σ maps W properly to $K^{\rho+k}(\alpha_1,\ldots,\alpha_{\rho+k-1},\beta)$. Hence σ maps $\sigma^{-1}(H) \cap W$ properly to H.

Since $\sigma^{-1}(H) \subset D$ (by (iii) and (iv) of Lemma (8.1.1)), $X \cap \sigma^{-1}(H)$ is a closed subset of $\sigma^{-1}(H)$. By (ii) of Lemma (8.1.1), $X \cap \sigma^{-1}(H) \subset W \cap \sigma^{-1}(H)$. Hence σ maps $\sigma^{-1}(H) \cap X$ properly to H.

Suppose Y is a branch of $Q \cap X$ and Y does not intersect $\sigma^{-1}(H) \cap X$. Because $Y \cap \sigma^{-1}(H) = \emptyset$ and because of (ii) of Lemma (8.1.1), Y is disjoint from the union of $\sigma^{-1}(H)$ and

$$K^{\rho+k-1}(\alpha_1,\ldots,\alpha_{\rho+k-1}) \times G^{n-\rho-k+1}(\alpha'_{\rho+k},\ldots,\alpha'_n;\alpha_{\rho+k},\ldots,\alpha_n) \cap \{|f| < \beta\}.$$

Hence Y is disjoint from

$$K^{\rho-1}(\alpha_1,\ldots,\alpha_{\rho-1}) \times G^{n-\rho+1}(\alpha'_\rho,\ldots,\alpha'_n;\alpha_\rho,\ldots,\alpha_n).$$

By Lemma (6.5)(b), $\dim Y \leqq \rho$, contradicting that $\dim Y = \rho + k > \rho$.

Since X has pure dimension $\rho + k$, (a) implies that σ makes $\sigma^{-1}(H) \cap X$ an analytic cover over H. The number of branches of $\sigma^{-1}(H) \cap X$ cannot exceed the number of sheets of $\sigma^{-1}(H) \cap X$. It follows from (b) that $Q \cap X$ has only a finite number of branches.

Q.E.D.

(8.1.3) Corollary. Suppose $X \cap Q$ is the restriction to D of a subvariety $\tilde{X}$ of pure dimension $\rho + k$ in Q. Then

(a) σ maps $\tilde{X}$ properly to $K^{\rho+k}(\alpha_1,\ldots,\alpha_{\rho+k-1},\beta)$, and

(b) $\sigma^{-1}(H) \cap \tilde{X} \subset X$.

Proof. By (ii) of Lemma (8.1.1), $\tilde{X}$ is disjoint from

$$K^{\rho+k-1}(\alpha_1,\ldots,\alpha_{\rho+k-1})\times G^{n-\rho-k+1}(\alpha'_{\rho+k},\ldots,\alpha'_n;\alpha_{\rho+k},\ldots,\alpha_n)\cap\{|f|<\beta\}.$$

Hence $\tilde{X}\subset W$, where W is the same as in the proof of Corollary (8.1.2). (a) follows, because σ maps W properly to $K^{\rho+k}(\alpha_1,\ldots,\alpha_{\rho+k-1},\beta)$. (b) follows from the fact that $\sigma^{-1}(H)\subset D$ ((iii) and (iv) of Lemma (8.1.1)). Q.E.D.

Since by (0.15) σ makes $\sigma^{-1}(H)\cap X$ an analytic cover over H whenever $\sigma^{-1}(H)\cap X\neq\emptyset$ ((a) of Corollary (8.12)) and the restriction map

$$\Gamma(K^{\rho+k}(\alpha_1,\ldots,\alpha_{\rho+k-1},\beta),{}_{\rho+k}\mathcal{O})\to\Gamma(H,{}_{\rho+k}\mathcal{O})$$

is bijective, by Proposition (7.2.6), $\sigma^{-1}(H)\cap X$ can be extended uniquely to a subvariety X* in Q which is an analytic cover over $K^{\rho+k}(\alpha_1,\ldots,\alpha_{\rho+k-1},\beta)$ under σ. $X^*\cap D = X\cap Q$, because every branch of $X^*\cap D$ and $X\cap Q$ intersects $\sigma^{-1}(H)\cap X$ ((b) of Corollary (8.1.2)). Therefore we have:

(8.1.4) <u>Corollary</u>. $X\cap Q$ can always be extended uniquely to a subvariety of pure dimension $\rho + k$ in Q.

Proof. Uniqueness follows, because (a) of Corollary (8.1.3) implies that a subvariety of pure dimension $\rho + k$ in Q which extends $X\cap Q$ is an analytic cover over $K^{\rho+k}(\alpha_1,\ldots,\alpha_{\rho+k-1},\beta)$. Q.E.D.

(8.2) Suppose f vanishes identically on X. Let $D' = D \cup (G-\{f=0\})$.

(8.2.1) Lemma. After a homogeneous affine transformation of the coordinates system of $\mathbb{C}^n$ there exist

$$\alpha_1 > 0,\ldots,\alpha_\rho > 0,\ \alpha_{\rho+1} > \alpha'_{\rho+1} > 0,\ldots,\alpha_n > \alpha'_n > 0$$

such that

(i) $K^n(\alpha_1,\ldots,\alpha_n) \subset G$,

(ii) $K^{\rho+k}(\alpha_1,\ldots,\alpha_{\rho+k}) \times G^{n-\rho-k}(\alpha'_{\rho+k+1},\ldots,\alpha'_n;\alpha_{\rho+k+1},\ldots,\alpha_n) \subset D' - X$,

(iii) $K^{\rho-1}(\alpha_1,\ldots,\alpha_{\rho-1}) \times G^{n-\rho+1}(\alpha'_\rho,\ldots,\alpha'_n;\alpha_\rho,\ldots,\alpha_n) \subset D'$, and

(iv) φ is regular in $z_1,\ldots,z_n$ on $K^n(\alpha_1,\ldots,\alpha_n)$.

Proof. Since f vanishes identically on X, X is a subvariety in D'.

As in the beginning of the proof of (8.1.1), after a homogeneous affine transformation and after a shrinking of G, we can assume that

(i) φ is regular in $z_1,\ldots,z_n$, and

(ii) $(z_1,\ldots,z_{\rho+k}): X \to \mathbb{C}^{\rho+k}$ has discrete fibers.

Let $E = \{z_1=\ldots=z_{\rho-1}=0\}$. Since $\varphi|E \cap G$ is strongly 1-convex and $f|E$ is the associated quadratic polynomial of $\varphi|E \cap G$ at 0, by

Lemma (6.2) we choose $\gamma > 0$ and an open neighborhood U of 0 in $E \cap G$ such that

(**) $$\varphi(z) \geqq \operatorname{Re} f(z) + \gamma \Sigma_{i=\rho}^{n} |z_i|^2$$

for $z \in U$.

Since $X \cap \{z_1 = \ldots = z_{\rho+k} = 0\}$ is discrete, we can choose

$$0 < \alpha'_{\rho+k+1} < \alpha_{\rho+k+1}, \ldots, 0 < \alpha'_n < \alpha_n$$

such that

(i) $\{z_1 = \ldots = z_{\rho+k} = 0,\ |z_{\rho+k+1}| \leqq \alpha_{\rho+k+1}, \ldots, |z_n| \leqq \alpha_n\} \subset U$, and

(ii) $A := \bigcup_{i=\rho+k+1}^{n} \{z_1 = \ldots = z_{\rho+k} = 0,\ \alpha'_i \leqq |z_i| \leqq \alpha_i,\ |z_j| \leqq \alpha_j$ for $\rho + k + 1 \leqq j \leqq n\}$ is disjoint from X.

By (**), $A \subset D'$. Since X is closed in D', $D' - X$ is an open neighborhood of A. We can choose positive numbers

$$\alpha_1^*, \ldots, \alpha_{\rho-1}^*,\ \alpha_\rho, \ldots, \alpha_{\rho+k}$$

such that

(i) $K^n(\alpha_1^*, \ldots, \alpha_{\rho-1}^*, \alpha_\rho, \ldots, \alpha_n) \subset G$,

(ii) $\{z_1 = \ldots = z_{\rho-1} = 0,\ |z_j| \leqq \alpha_j$ for $\rho \leqq j \leqq n\} \subset U$, and

(iii) $\bigcup_{i=\rho+k+1}^{n} \{|z_1| < \alpha_1^*, \ldots, |z_{\rho-1}| < \alpha_{\rho-1}^*,\ |z_\rho| < \alpha_\rho, \ldots, |z_{\rho+k}| < \alpha_{\rho+k},$ $\alpha'_i \leqq |z_i| \leqq \alpha_i,\ |z_j| \leqq \alpha_j$ for $\rho+k+1 \leqq j \leqq n\}$ is contained in $D' - X$.

Choose $0 < \alpha_i' < \alpha_i$ for $\rho \leqq i \leqq \rho + k$. By (**),

$$\bigcup_{i=\rho}^{n}\{z_1=\ldots=z_{\rho-1}=0,\ \alpha_i' \leqq |z_i| \leqq \alpha_i,\ |z_j| \leqq \alpha_j \text{ for } \rho \leqq j \leqq n\} \subset D'.$$

Hence we can choose

$$0 < \alpha_1 < \alpha_1^* , \ldots, 0 < \alpha_{\rho-1} < \alpha_{\rho-1}^*$$

such that condition (iii) of the statement of the lemma is satisfied.

Q.E.D.

(8.2.2) <u>Corollary</u>. Let $Q' = K^n(\alpha_1,\ldots,\alpha_n)$ and $\sigma': Q' \to \mathbb{C}^{\rho+k}$ be defined by $z_1,\ldots,z_{\rho+k}$. Let

$$H' = K^{\rho-1}(\alpha_1,\ldots,\alpha_{\rho-1}) \times G^{k+1}(\alpha_\rho',\ldots,\alpha_{\rho+k}';\alpha_\rho,\ldots,\alpha_{\rho+k}).$$

Then

(a) σ' maps $(\sigma')^{-1}(H') \cap X$ properly to H', and

(b) every branch of $Q' \cap X$ intersects $(\sigma')^{-1}(H') \cap X$.

Hence $Q' \cap X$ has only a finite number of branches.

Proof. Let $W' = K^{\rho+k}(\alpha_1,\ldots,\alpha_{\rho+k}) \times \overline{K}^{n-\rho-k}(\alpha_{\rho+k+1}',\ldots,\alpha_n')$. σ' maps W' properly to $K^{\rho+k}(\alpha_1,\ldots,\alpha_{\rho+k})$. Hence σ' maps $(\sigma')^{-1}(H') \cap W'$ properly to H'.

Since $(\sigma')^{-1}(H') \subset D'$ ((iii) of Lemma (8.2.1)), $(\sigma')^{-1}(H') \cap X$

is a closed subset of $(\sigma')^{-1}(H')$. By (ii) of Lemma (8.2.1), $(\sigma')^{-1}(H') \cap X \subset (\sigma')^{-1}(H') \cap W'$. Hence σ' maps $(\sigma')^{-1}(H') \cap X$ properly to H'.

Suppose Y is a branch of $Q' \cap X$ which does not intersect $(\sigma')^{-1}(H') \cap X$. Because $Y \cap (\sigma')^{-1}(H') = \emptyset$ and because of (ii) of Lemma (8.2.1), Y is disjoint from the union of $(\sigma')^{-1}(H')$ and

$$K^{\rho+k}(\alpha_1,\ldots,\alpha_{\rho+k}) \times G^{n-\rho-k}(\alpha'_{\rho+k+1},\ldots,\alpha'_n;\alpha_{\rho+k+1},\ldots,\alpha_n).$$

Hence Y is disjoint from

$$K^{\rho-1}(\alpha_1,\ldots,\alpha_{\rho-1}) \times G^{n-\rho+1}(\alpha'_\rho,\ldots,\alpha'_n;\alpha_\rho,\ldots,\alpha_n).$$

By Lemma (6.5)(a), $\dim Y \leqq \rho - 1$, contradicting that $\dim Y = \rho + k > \rho - 1$.

Since X has pure dimension $\rho + k$, (a) implies that σ' makes $(\sigma')^{-1}(H') \cap X$ an analytic cover over H'. The number of branches of $(\sigma')^{-1}(H') \cap X$ cannot exceed the number of sheets of $(\sigma')^{-1}(H') \cap X$. It follows from (b) that $Q' \cap X$ has only a finite number of branches.

Q.E.D.

(8.2.3) Corollary. Suppose $Q' \cap X$ is the restriction to D of a subvariety $\tilde{X}$ of pure dimension $\rho + k$ in Q'. Then

(a) σ' maps $\tilde{X}$ properly to $K^{\rho+k}(\alpha_1,\ldots,\alpha_{\rho+k})$, and

(b) $(\sigma')^{-1}(H') \cap \tilde{X} \subset X$.

Proof. We first show that f vanishes identically on $\tilde{X}$. It suffices to prove that every branch of $\tilde{X}$ intersects D. Suppose some branch X' of $\tilde{X}$ does not intersect D. Then X": = X' $\cap$ {f=0} is disjoint from D'. By (iii) of Lemma (8.2.1), X" is disjoint from

$$K^{\rho-1}(\alpha_1,\ldots,\alpha_{\rho-1}) \times G^{n-\rho+1}(\alpha'_\rho,\ldots,\alpha'_n;\alpha_\rho,\ldots,\alpha_n).$$

For $x \in X''$, $\{z_1=z_1(x),\ldots,z_{\rho-1}=z_{\rho-1}(x)\} \cap X''$ is compact and hence is at most 0-dimensional. Therefore $\dim X'' \leqq \rho - 1$ and $\dim X' \leqq \rho$, contradicting $\dim X' = \rho + k > \rho$.

Since f vanishes identically on $\tilde{X}$, we have $Q' \cap X = D' \cap \tilde{X}$. By (ii) of Lemma (8.2.1), $\tilde{X}$ is disjoint from

$$K^{\rho+k}(\alpha_1,\ldots,\alpha_{\rho+k}) \times G^{n-\rho-k}(\alpha'_{\rho+k+1},\ldots,\alpha'_n;\alpha_{\rho+k+1},\ldots,\alpha_n).$$

Hence $\tilde{X} \subset W'$, where W' is the same as in the proof of Corollary (8.2.2). (a) follows, because σ' maps W' properly to $K^{\rho+k}(\alpha_1,\ldots,\alpha_{\rho+k})$. (b) follows from the fact that $(\sigma')^{-1}(H') \subset D'$ ((iii) of Lemma (8.2.1)). Q.E.D.

Since by (0.15) σ' makes $(\sigma')^{-1}(H') \cap X$ an analytic cover over H' whenever $(\sigma')^{-1}(H') \cap X \neq \emptyset$ ((i) of Corollary (8.2.2)) and the restriction map

$$\Gamma(K^{\rho+k}(\alpha_1,\ldots,\alpha_{\rho+k}), {}_{\rho+k}\mathcal{O}) \to \Gamma(H', {}_{\rho+k}\mathcal{O})$$

is bijective, by Proposition (7.2.6), $(\sigma')^{-1}(H') \cap X$ can be extended uniquely to a subvariety X^* in Q' which is an analytic cover over $K^{\rho+k}(\alpha_1,\ldots,\alpha_{\rho+k})$ under σ'. f vanishes identically on X^*, because every branch of X^* intersects $(\sigma')^{-1}(H') \cap X$. $X^* \cap D = X \cap Q'$, because every branch of $X^* \cap D$ and $X \cap Q'$ intersects $(\sigma')^{-1}(H') \cap X$ ((b) of Corollary (8.2.2)). Therefore we have:

(8.2.4) Corollary. $X \cap Q'$ can always be extended uniquely to a subvariety of pure dimension $\rho + k$ in Q'.

Proof. Uniqueness follows, because (a) of Corollary (8.2.3) implies that a subvariety of pure dimension $\rho + k$ in Q' which extends $X \cap Q'$ is an analytic cover over $K^{\rho+k}(\alpha_1,\ldots,\alpha_{\rho+k})$. Q.E.D.

(8.3) Theorem(subvariety extension). Suppose X is a complex space and D is an open subset of X which is ρ-concave at a point x of X. Suppose V is a subvariety in D and every branch of V has dimension $\geqq \rho + 1$. Then

(a) there exist an open neighborhood U of x in X and a subvariety $\tilde{V}$ in $D \cup U$ such that $\tilde{V} \cap D = V$;

(b) if every branch-germ of $\tilde{V}$ at x has dimension $\geqq \rho$, then the germ of $\tilde{V}$ at x is unique and every branch-germ of $\tilde{V}$ at x has dimension $\geqq \rho + 1$;

(c) U can be chosen such that only a finite number of branches of V intersect U.

Proof. W.l.o.g. we can assume the following:

(i) X is an open subset of some $\mathbb{C}^n$,

(ii) $x = 0$, and

(iii) $D = \{\varphi > 0\}$ for some strongly ρ-convex function φ on X with $\varphi(0) = 0$.

Let f be the associated quadratic polynomial of φ at 0. Let V_ℓ be the union of all ℓ-dimensional branches of V where f does not vanish identically. Let V_ℓ' be the union of all ℓ-dimensional branches of V where f vanishes identically.

By Corollary (8.1.4) we can find an open neighborhood Q_ℓ of 0 in X such that $V_\ell \cap Q_\ell$ can be extended to a subvariety $\tilde{V}_\ell$ of pure dimension ℓ in Q_ℓ. By Corollary (8.2.4) we can find an open neighborhood Q_ℓ' of 0 in X such that $V_\ell' \cap Q_\ell'$ can be extended to a subvariety $\tilde{V}_\ell'$ of pure dimension ℓ in Q_ℓ'.

Let $U = \bigcap_{\ell=\rho+1}^{n} (Q_\ell \cap Q_\ell')$ and let

$$\tilde{V} = V \cup \left(\bigcup_{\ell=\rho+1}^{n} U \cap (\tilde{V}_\ell \cup \tilde{V}_\ell')\right).$$

Then $\tilde{V}$ is a subvariety in $D \cup U$ and every branch of $\tilde{V}$ has dimension $\geqq \rho + 1$. (a) is proved.

(b) follows from Proposition (6.8) and the fact that every branch-germ of $\tilde{V}$ at x has dimension $\geqq \rho + 1$.

By virtue of Corollaries (8.1.2) and (8.2.2), only a finite number of branches of V_ℓ intersect Q_ℓ and only a finite number of V_ℓ' intersect Q_ℓ. Hence only a finite number of branches of V intersect U.

Q.E.D.

(8.4) Theorem (projection lemma). Suppose X is a complex space of pure dimension $\rho + k$ (where $k \geqq 1$) and D is an open subset of X which if strongly ρ-concave at a point x of X. Then, for some open neighborhood U of x in X, the branches of U can be divided into 2 groups whose unions are X_1 and X_2:

$$U = X_1 \cup X_2,$$

such that the following condition is satisfied. If $x \in X_i$, then there exist

(i) $\alpha_1^{(i)} > 0, \ldots, \alpha_{\rho-1}^{(i)} > 0$, $\alpha_\rho^{(i)} > \beta_\rho^{(i)} > 0, \ldots, \alpha_{\rho+k}^{(i)} > \beta_{\rho+k}^{(i)} > 0$,

(ii) an open neighborhood U_i of x in X_i, and

(iii) a proper holomorphic map with finite fibers $\pi_i : U_i \to K^{\rho+k}(\alpha_1^{(i)}, \ldots, \alpha_{\rho+k}^{(i)})$

such that

(i) π_i is the restriction of a holomorphic map from U to $\mathbb{C}^{\rho+k}$,

(ii) $\pi_i^{-1}(H_i) \subset D$, and

(iii) every branch of $U_i \cap D$ intersects $\pi_i^{-1}(H_i)$,

where H_i is the union of

$$K^{\rho}(\alpha_1^{(1)},\ldots,\alpha_\rho^{(1)}) \times G^k(\beta_{\rho+1}^{(1)},\ldots,\beta_{\rho+k}^{(1)};\alpha_{\rho+1}^{(1)},\ldots,\alpha_{\rho+k}^{(1)})$$

and

$$K^{\rho+k}(\alpha_1^{(1)},\ldots,\alpha_{\rho+k}^{(1)}) \cap \{|z_\rho - \beta_\rho^{(1)}| < \frac{\beta_\rho^{(1)}}{2}\},$$

and

$$H_2 = K^{\rho-1}(\alpha_1^{(2)},\ldots,\alpha_{\rho-1}^{(2)}) \times G^{k+1}(\beta_\rho^{(2)},\ldots,\beta_{\rho+k}^{(2)};\alpha_\rho^{(2)},\ldots,\alpha_{\rho+k}^{(2)}).$$

Proof. Since the problem is local in nature, we can assume w.l.o.g. the following:

(i) X is a complex subspace of an open neighborhood G of 0 in some $\mathbb{C}^n$,

(ii) $x = 0$, and

(iii) $D = X \cap \{\varphi > 0\}$, where φ is a strongly ρ-convex function on G with $\varphi(0) = 0$.

Let f be the associated quadratic polynomial of φ at 0. Let X_1 be the union of branches of X where f does not vanish identically. Let X_2 be the union of branches of X where f vanishes identically.

By Corollaries (8.1.2), (8.1.3), (8.2.2), and (8.2.3) we can find

$$\alpha_1^{(i)} > 0,\ldots,\alpha_{\rho-1}^{(i)} > 0,\ \alpha_\rho^{(i)} > \beta_\rho^{(i)} > 0,\ldots,\alpha_n^{(i)} > \beta_n^{(i)} > 0$$

and a proper holomorphic map

$$\sigma_i : X_i \cap Q_i \to K^{\rho+k}(\alpha_1^{(i)},\ldots,\alpha_{\rho+k}^{(i)}),$$

where $Q_i = K^n(\alpha_1^{(i)},\ldots,\alpha_n^{(i)})$, such that

(i) $Q_i \subset G$,

(ii) σ_1 is defined by f and $\rho + k - 1$ linear functions,

(iii) σ_2 is defined by $\rho + k$ linear functions,

(iv) $\sigma_i^{-1}(H_i) \subset D$, and

(v) every branch of $X_i \cap Q_i \cap D$ intersects $\sigma_i^{-1}(H_i)$.

To finish the proof, we need only set $U = X$, $U_i = X_i \cap Q_i$ and, $\pi_i = \sigma_i | U_i$ (when $x \in X_i$). Q.E.D.

§9 Subsheaf extension.

(9.1) Lemma. Suppose G is a connected open subset of $\mathbb{C}^n$ and $\mathcal{F}$ is a coherent analytic sheaf on G. Suppose φ: ${}_n\mathcal{O}^r \to \mathcal{F}$ is a sheaf-homomorphism on G such that φ is an isomorphism at some point of G. Then φ is injective and there exists a unique sheaf-homomorphism ψ: $\mathcal{F} \to \mathfrak{M}^r$ (where $\mathfrak{M}$ is the sheaf of germs of meromorphic functions on $\mathbb{C}^n$) such that

$$\begin{array}{ccc} {}_n\mathcal{O}^r & \xrightarrow{\ \varphi\ } & \mathcal{F} \\ \Big\downarrow i & \swarrow \psi & \\ \mathfrak{M}^r & & \end{array}$$

is commutative. Moreover, ψ is injective if $\mathcal{F}$ is torsion-free.

Proof. Since φ is an isomorphism at some point of G, Supp Ker φ and Supp Coker φ are both subvarieties of codimension ≥ 1 in G. Hence Ker $\varphi = 0$. φ is injective.

Let X = Supp Coker φ. Let $\mathcal{I}$ be the conductor sheaf from $\mathcal{F}$ to Im φ, i.e.

$$\mathcal{I}_x = \{s \in {}_n\mathcal{O}_x \mid s\,\mathcal{F}_x \subset \mathrm{Im}\,\varphi_x\}.$$

$\mathcal{I}$ is coherent and the zero set of $\mathcal{I}$ is X.

First, we show the uniqueness of ψ. Suppose ψ' is another sheaf-homomorphism satisfying the requirements. Then $\psi = \psi'$ on G - X, because both are equal to $i\varphi^{-1}|G-X$. Supp Im$(\psi - \psi') \subset X$. Since non-zero sections of $\mathfrak{M}^r$ cannot have thin supports, Im$(\psi-\psi') = 0$. $\psi=\psi'$.

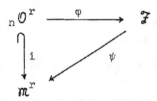

Since we have the uniqueness of ψ, to show its existence, we need only do it locally. Hence we can assume w.l.o.g. that there exists $0 \neq f \in \Gamma(G, \mathcal{J})$. For $x \in G$ and $s \in \mathcal{F}_x$, define $\psi(s) = f^{-1}\varphi^{-1}(fs)$. ψ satisfies the requirement.

Suppose $\mathcal{F}$ is torsion-free. ψ is clearly injective on $G - X$, because ψ agrees with $i\,\varphi^{-1}$ on $G - X$. $\operatorname{Ker}\psi \subset X$. It follows that $\operatorname{Ker}\psi = 0$. ψ is injective. Q.E.D.

(9.2) Suppose $G \subset \tilde{G}$ are connected open subsets of $\mathbb{C}^n$. Suppose $\mathcal{A}$ is a coherent analytic subsheaf of ${}_n\mathcal{O}^p|G$. Let $\mathcal{F} = {}_n\mathcal{O}^p/\mathcal{A}$. Let $s_i \in \Gamma(G, \mathcal{F})$ be the image of

$$(0,\ldots,0,1,0,\ldots,0) \in \Gamma(G, {}_n\mathcal{O}^p)$$

under the natural sheaf-epimorphism $\eta: {}_n\mathcal{O}^p \to \mathcal{F}$ on G, where 1 is in the ith position.

(9.2.1) <u>Lemma</u>. $\mathcal{A}$ can be extended coherently to $\tilde{G}$ as a subsheaf of ${}_n\mathcal{O}^p$ if and only if

(i) $\mathcal{F}$ can be extended to a coherent analytic sheaf $\tilde{\mathcal{F}}$ on $\tilde{G}$, and

(ii) s_i can be extended to some element $\tilde{s}_i$ of $\Gamma(\tilde{G}, \tilde{\mathcal{F}})$.

Proof. Suppose $\mathcal{A}$ can be extended to a coherent analytic subsheaf $\tilde{\mathcal{A}}$ of ${}_n\mathcal{O}^p$ on $\tilde{G}$. Let $\tilde{\mathcal{F}} = {}_n\mathcal{O}^p/\tilde{\mathcal{A}}$ and let $\tilde{\eta}: {}_n\mathcal{O}^p \to \tilde{\mathcal{F}}$ be the natural sheaf-epimorphism. Let $\tilde{s}_i \in \Gamma(\tilde{G}, \tilde{\mathcal{F}})$ be the image of

$$(0,\dots,0,1,0,\dots,0) \in \Gamma(\tilde{G}, {}_n\mathcal{O}^p)$$

under η, where 1 is in the ith position. Then $\tilde{\mathfrak{F}}$ extends $\mathfrak{F}$ and $\tilde{s}_i$ extends s_i.

Conversely, suppose $\tilde{\mathfrak{F}}$ is a coherent analytic sheaf on $\tilde{G}$ extending $\mathfrak{F}$ and $\tilde{s}_i \in \Gamma(\tilde{G}, \tilde{\mathfrak{F}})$ extends s_i. Let $\tilde{\eta}$: ${}_n\mathcal{O}^p \to \tilde{\mathfrak{F}}$ be defined by the p sections $\tilde{s}_1,\dots,\tilde{s}_p$ (i.e. for $x \in \tilde{G}$ and $a_1,\dots,a_p \in {}_n\mathcal{O}_x$, $\tilde{\eta}(a_1,\dots,a_p) = \Sigma_{i=1}^p a_i(\tilde{s}_i)_x$). Then Ker $\tilde{\eta}$ is a coherent analytic subsheaf of ${}_n\mathcal{O}^p|\tilde{G}$ extending $\mathcal{S}$. Q.E.D.

Let $S = S_{n-1}(\mathfrak{F})$. S is a subvariety of dimension $\leq n - 1$ in G. Fix $x \in G - S$. For some open neighborhood U of x in G - S and for some r, there is a sheaf-isomorphism σ: $\mathfrak{F}|U \to {}_n\mathcal{O}^r|U$. Since $\sigma\eta$: ${}_n\mathcal{O}^p|U \to {}_n\mathcal{O}^r|U$ is surjective, at every point of U the $r \times p$ matrix of holomorphic functions on U representing $\sigma\eta$ has rank r. Hence we can select $s_{i_1},\dots,s_{i_r}$ such that the $r \times r$ matrix of holomorphic functions with $\sigma(s_{i_1}),\dots,\sigma(s_{i_r})$ as columns has rank r at x. Let φ: ${}_n\mathcal{O}^r \to \mathfrak{F}$ be the sheaf-homomorphism on G defined by $s_{i_1},\dots,s_{i_r}$. Then φ is an isomorphism at x. By Lemma (9.1) there exists a unique sheaf-homomorphism ψ: $\mathfrak{F} \to \mathfrak{M}^r$ such that

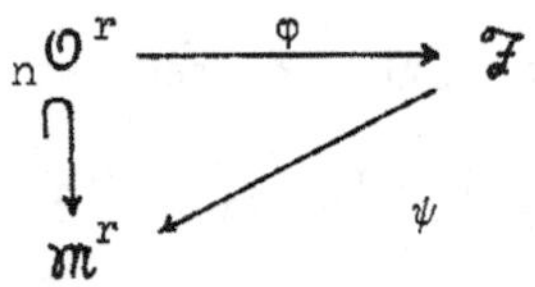

is commutative. Let $t_i = \psi(s_i)$. We call $t_1,\ldots,t_p$ a set of <u>associated meromorphic vector-functions</u> for $\mathcal{A}$.

(9.2.2) <u>Proposition.</u> Suppose $\mathcal{A}_{n-1} = \mathcal{A}$. Then $\mathcal{A}$ can be extended coherently to $\tilde{G}$ as a subsheaf of ${}_n\mathcal{O}^p$ if and only if $t_1,\ldots,t_p$ can be extended to r-tuples of meromorphic functions on $\tilde{G}$.

Proof. If $\mathcal{A}$ can be extended to a coherent analytic subsheaf $\tilde{\mathcal{A}}$ of ${}_n\mathcal{O}^p|\tilde{G}$, then we can assume $\tilde{\mathcal{A}}_{n-1} = \tilde{\mathcal{A}}$. We repeat the preceding argument with $\tilde{\mathcal{F}} = {}_n\mathcal{O}^p/\tilde{\mathcal{A}}$ instead of $\mathcal{F}$ (using the same $i_1,\ldots,i_r$, the same x and U) and obtain associated meromorphic vector-functions $\tilde{t}_1,\ldots,\tilde{t}_p$ for $\tilde{\mathcal{A}}$. $\tilde{t}_i$ is an r-tuple of meromorphic functions on $\tilde{G}$ extending t_i $(1 \leq i \leq p)$.

Conversely, if t_i can be extended to an r-tuple t_i^* of meromorphic functions on $\tilde{G}$ $(1 \leq i \leq p)$, then the subsheaf $\mathcal{F}^*$ of $\mathcal{M}^r|G$ generated by $t_1^*,\ldots,t_p^*$ is coherent. For, if f is a non-identically-zero holomorphic functions on some connected open subset D of $\tilde{G}$ such that f t_i^* is an r-tuple of holomorphic functions on D, then $\mathcal{F}^* \approx f\,\mathcal{F}^*$ on D and f $\mathcal{F}^*$ is the subsheaf of ${}_n\mathcal{O}^r|D$ generated by $ft_1^*,\ldots,ft_p^*$. Since $\mathcal{A}_{n-1} = \mathcal{A}$, $\mathcal{F}$ is torsion-free. ψ is therefore injective and we can identify $\mathcal{F}$ with $\psi(\mathcal{F})$. Then $\mathcal{F}^*$ extends $\mathcal{F}$ and t_i^* extends s_i $(1 \leq i \leq p)$. By Lemma (9.2), $\mathcal{A}$ can be extended. Q.E.D.

(9.2.3) <u>Proposition.</u> Suppose $\tilde{G}$ is Stein and the restriction map $\Gamma(\tilde{G},\mathcal{M}) \rightarrow \Gamma(G,\mathcal{M})$ is bijective. Suppose $\mathcal{G}$ is a coherent analytic sheaf on $\tilde{G}$. If $\mathcal{R}$ is a coherent analytic subsheaf of $\mathcal{G}|G$ and $\mathcal{R}_{n-1} = \mathcal{R}$, then $\mathcal{R}$ can be extended to a coherent analytic subsheaf of $\mathcal{G}$.

Proof. We consider first the special case where there is a sheaf-epimorphism $\lambda: {}_n\mathcal{O}^p \to \mathcal{G}$ on $\tilde{G}$. Then $\lambda^{-1}(\mathcal{R})$ is a coherent analytic subsheaf of ${}_n\mathcal{O}^p|G$ and $\lambda^{-1}(\mathcal{R})_{n-1} = \lambda^{-1}(\mathcal{R})$. Since the restriction map $\Gamma(\tilde{G},\mathcal{M}) \to \Gamma(G,\mathcal{M})$ is bijective, $\lambda^{-1}(\mathcal{R})$ can be extended to a coherent analytic subsheaf $\mathcal{J}$ of ${}_n\mathcal{O}^p|\tilde{G}$. $\lambda(\mathcal{J})$ is a coherent analytic subsheaf of $\mathcal{G}$ extending $\mathcal{R}$. The special case is proved.

For the general case, we observe that, since $\tilde{G}$ is Stein, $\mathcal{G}$ is generated by a countable number of global sections. Hence we can find an increasing sequence of coherent analytic subsheaves of $\mathcal{G}$

$$\mathcal{G}^{(1)} \subset \mathcal{G}^{(2)} \subset \ldots \subset \mathcal{G}^{(k)} \subset \mathcal{G}^{(k+1)} \subset \ldots \subset \mathcal{G}$$

such that $\mathcal{G} = \bigcup_{k=1}^{\infty} \mathcal{G}^{(k)}$ and $\mathcal{G}^{(k)}$ is generated by a finite number of global sections. There exists a sheaf-epimorphism $\lambda_k: {}_n\mathcal{O}^{p_k} \to \mathcal{G}^{(k)}$ on $\tilde{G}$. Let $\mathcal{R}^{(k)} = \mathcal{R} \cap \mathcal{G}^{(k)}$. Then $\mathcal{R}^{(k)}$ is a coherent analytic subsheaf of $\mathcal{G}^{(k)}|G$ such that $(\mathcal{R}^{(k)})_{n-1} = \mathcal{R}^{(k)}$. Therefore $\mathcal{R}^{(k)}$ can be extended to a coherent analytic subsheaf $\tilde{\mathcal{R}}^{(k)}$ of $\mathcal{G}^{(k)}$. Let $\tilde{\mathcal{R}} = \Sigma_{k=1}^{\infty} \tilde{\mathcal{R}}^{(k)}$. Then $\tilde{\mathcal{R}}$ is a coherent analytic subsheaf of $\mathcal{G}$ extending $\mathcal{R}$. Q.E.D.

(9.3) <u>Theorem</u> (subsheaf extension). Suppose $(X, \mathcal{O})$ is a complex space and D is an open subset of X which is strongly ρ-concave at a point x_0 of X. Suppose $\mathcal{G}$ is a coherent analytic sheaf on X and $\mathcal{F}$

is a coherent analytic subsheaf of $\mathcal{G}|D$. If $\mathcal{F}_\rho = \mathcal{F}$, then $\mathcal{F}$ can be extended coherently to an open neighborhood of x_0 as a subsheaf of $\mathcal{G}$.

Proof. (a) We prove by descending induction on ρ. Let $n = \dim_x X$. When $\rho \geqq n$, $\mathcal{F}$ agrees with $\mathcal{G}$ on an open neighborhood of x_0 (because $\mathcal{F}_\rho = \mathcal{F}$). In that case, the extendability of $\mathcal{F}$ is trivial.

Suppose $\rho < n$. By induction hypothesis, $\mathcal{F}_{\rho+1}$ can be extended coherently to an open neighborhood of x_0 as a subsheaf of $\mathcal{G}$. By replacing X by an open neighborhood of x_0 and by replacing $\mathcal{G}$ by $\mathcal{F}_{\rho+1}$, we can assume w.l.o.g. that $\mathcal{F}_{\rho+1} = \mathcal{G}$.

(b) Let $Y = \operatorname{Supp} \mathcal{G}/\mathcal{F}$. We can assume that $Y \neq \emptyset$. Since $\mathcal{F}_{\rho+1} = \mathcal{G}$ and $\mathcal{F}_\rho = \mathcal{F}$, Y has pure dimension $\rho + 1$ in D. By Theorem (8.3) we can find an open neighborhood W of x_0 in X such that

(i) Y can be extended to a subvariety $\tilde{Y}$ of pure dimension $\rho + 1$ in $D \cup W$, and

(ii) only a finite number of branches of Y intersect W.

Let $\{Y_i\}_{i \in I}$ be the set of branches of Y which intersect W. Take $x_i \in Y_i$. Let $\mathcal{J}$ be the ideal-sheaf of $\tilde{Y}$. Let $\mathcal{I}$ be the conductor sheaf from $\mathcal{G}$ into $\mathcal{F}$, i.e.

$$\mathcal{I}_x = \{s \in \mathcal{O}_x | s\, \mathcal{G}_x \subset \mathcal{F}_x\} \text{ for } x \in D.$$

Since the zero set of $\mathcal{J}$ is Y, by Hilbert Nullstellensatz we can find a positive integer k_i such that $\mathcal{J}_{x_i}^{k_i} \subset \mathcal{J}_{x_i}$. Let $k = \max_{i \in I} k_i$. Then $(\mathcal{J}^k \mathcal{G})_{x_i} \subset \mathcal{F}_{x_i}$ for $i \in I$. Let Z be the subvariety of D where $\mathcal{J}^k \mathcal{G}$ is not contained in $\mathcal{F}$. Since $\mathcal{F}_\rho = \mathcal{F}$, Z is either empty or a subvariety of pure dimension $\rho + 1$ in Y. Since $x_i \notin Z$, Z contains no Y_i. Hence $Z \cap W = \emptyset$.

By replacing X by W, we can assume that X = W. Then $\mathcal{J}^k \mathcal{G} \subset \mathcal{F}$.

Give $\tilde{Y}$ the structure sheaf $(\mathcal{O}/\mathcal{J}^k)|\tilde{Y}$. Let $\mathcal{F}' = \mathcal{F}/\mathcal{J}^k \mathcal{G}$ and $\mathcal{G}' = \mathcal{G}/\mathcal{J}^k \mathcal{G}$. $\mathcal{G}'$ (when restricted to $\tilde{Y}$) can be regarded as a coherent analytic sheaf on the complex space $\tilde{Y}$. $\mathcal{F}'$ is a coherent analytic sheaf of $\mathcal{G}'$ on Y. Let λ: $\mathcal{G} \to \mathcal{G}'$ be the natural sheaf-epimorphism. Then $\mathcal{F} = \lambda^{-1}(\mathcal{F}')$. Hence $\mathcal{F}$ can be extended coherently to an open neighborhood of x_0 as a subsheaf of $\mathcal{G}$ if $\mathcal{F}'$ can be extended coherently to an open neighborhood of x_0 as a subsheaf of $\mathcal{G}'$.

By replacing X by $\tilde{Y}$, $\mathcal{F}$ by $\mathcal{F}'$, and $\mathcal{G}$ by $\mathcal{G}'$, we can assume w.l.o.g. that $X = \tilde{Y}$ has pure dimension $\rho + 1$.

(c) Apply Theorem (8.4) to X. For some open neighborhood U of x_0 in X, the branches of U can be divided into 2 groups whose unions are X_1 and X_2:

$$U = X_1 \cup X_2 ,$$

such that the following condition is satisfied. If $x_0 \in X_i$, then there exist

(i) $\alpha_1^{(i)} > 0, \ldots, \alpha_{\rho-1}^{(i)} > 0$, $\alpha_\rho^{(i)} > \beta_\rho^{(i)} > 0$, $\alpha_{\rho+1}^{(i)} > \beta_{\rho+1}^{(i)} > 0$,

(ii) an open neighborhood U_i of x_0 in X_i, and

(iii) a proper holomorphic map with finite fibers

$\pi_i: U_i \to K^{\rho+1}(\alpha_1^{(i)}, \ldots, \alpha_{\rho+1}^{(i)})$ such that

(i) π_i is the restriction of a holomorphic map from U to $\mathbb{C}^{\rho+1}$,

(ii) $\pi_i^{-1}(H_i) \subset D$, and

(iii) every branch of $U_i \cap D$ intersects $\pi_i^{-1}(H_i)$,

where H_1 is the union of

$$K^\rho(\alpha_1^{(1)}, \ldots, \alpha_\rho^{(1)}) \times G^1(\beta_{\rho+1}^{(1)}; \alpha_{\rho+1}^{(1)})$$

and

$$K^{\rho+1}(\alpha_1^{(1)}, \ldots, \alpha_{\rho+1}^{(1)}) \cap \{|z_\rho - \beta_\rho^{(1)}| < \frac{\beta_\rho^{(1)}}{2}\},$$

and

$$H_2 = K^{\rho-1}(\alpha_1^{(2)}, \ldots, \alpha_{\rho-1}^{(2)}) \times G^2(\beta_\rho^{(2)}, \beta_{\rho+1}^{(2)}; \alpha_\rho^{(2)}, \alpha_{\rho+1}^{(2)}).$$

By replacing X by U, we can assume that X = U. Since $\mathfrak{F}_\rho = \mathfrak{F}$, we have $\mathfrak{F} = \mathfrak{F}[X_1] \cap \mathfrak{F}[X_2]$. To extend $\mathfrak{F}$, we need only extend both $\mathfrak{F}[X_1]$ and $\mathfrak{F}[X_2]$. We shall only prove the extendibility of

$\mathfrak{F}[X_1]$. The extendibility of $\mathfrak{F}[X_2]$ can be proved in an entirely similar way. Supp $\mathfrak{G}/\mathfrak{F}[X_1] = D \cap X_2$. By replacing $\mathfrak{F}$ by $\mathfrak{F}[X_1]$ and repeating the argument in (b), we can assume that $X = X_2 = U_2$. We have only U_2, $\alpha_i^{(2)}$, $\beta_i^{(2)}$, π_2, and H_2. We shall drop the index 2 from these notations.

(d) $\pi_0(\mathfrak{G})$ is a coherent analytic sheaf on $K^{\rho+1}(\alpha_1,\ldots,\alpha_{\rho+1})$ and $\pi_0(\mathfrak{F}|\pi^{-1}(H))$ is a coherent analytic subsheaf of $\pi_0(\mathfrak{G})|H$, because π is proper and has finite fibers (see (0.20)). Since π has finite fibers, we have

$$\pi_0(\mathfrak{F}|\pi^{-1}(H))_\rho = \pi_0(\mathfrak{F}|\pi^{-1}(H)).$$

Since every meromorphic function on H can be uniquely extended to $K^{\rho+1}(\alpha_1,\ldots,\alpha_{\rho+1})$, by Proposition (8.2.3), $\pi_0(\mathfrak{F}|\pi^{-1}(H))$ can be extended coherently to $K^{\rho+1}(\alpha_1,\ldots,\alpha_{\rho+1})$ as a subsheaf of $\pi_0(\mathfrak{G})$. $\pi_0(\mathfrak{F}|\pi^{-1}(H))$ can be generated by global sections of $\pi_0(\mathfrak{G})$. Hence $\mathfrak{F}|\pi^{-1}(H)$ can be generated by global sections of $\mathfrak{G}$. $\mathfrak{F}|\pi^{-1}(H)$ can be extended to a coherent analytic subsheaf $\mathfrak{F}^*$ of $\mathfrak{G}$. We can assume that $\mathfrak{F}^*_\rho = \mathfrak{F}^*$. Let A be the set of points in D where $\mathfrak{F}^*$ and $\mathfrak{F}$ disagree. Since $\mathfrak{F}^*_\rho = \mathfrak{F}^*$ and $\mathfrak{F}_\rho = \mathfrak{F}$, A is either empty or has pure dimension $\rho + 1$. Since $A \cap \pi^{-1}(H) = \emptyset$ and every branch of D intersects $\pi^{-1}(H)$, A must be empty. $\mathfrak{F}^*$ extends $\mathfrak{F}$. Q.E.D.

§10 Globalization

(10.1) Lemma. For any natural number m, $\varphi = (\Sigma_{i=1}^n |z_i|^{2m})^{-1}$ is strongly n-convex on $\mathbb{C}^n - \{0\}$.

Proof. For $a_1,\dots,a_n \in \mathbb{C}$

$$\Sigma_{i,j=1}^n \frac{\partial^2\varphi}{\partial z_i \partial \bar{z}_j} a_i\bar{a}_j = \varphi^3(2m^2|\Sigma_{i=1}^n z_i^{m-1}\bar{z}_i^m a_i|^2 -$$

$$m^2(\Sigma_{i=1}^n z_i^{m-1}\,\bar{z}_i^{m-1}\, a_i\bar{a}_i)\varphi^{-1}).$$

Hence, when $a_i = z_i (1 \leqq i \leqq n)$ and $(z_1,\dots,z_n) \neq 0$,

$$\Sigma_{i,j=1}^n \frac{\partial^2\varphi}{\partial z_i \partial \bar{z}_j} a_i\bar{a}_j = m^2\varphi > 0 .$$

At every point of $\mathbb{C}^n - \{0\}$, the hermitian matrix $(\frac{\partial^2\varphi}{\partial z_i \partial \bar{z}_j})$ has at least one positive eigenvalue. φ is strongly n-convex on $\mathbb{C}^n - \{0\}$.

Q.E.D.

(10.2) Denote the coordinates of $\mathbb{C}^{\rho+k}$ by $z_1,\dots,z_\rho, w_1,\dots,w_k$. Fix a natural number m. For $0 < \tau < 1$, let R_τ denote the following domain in $\mathbb{C}^{\rho+k}$:

$$R_\tau = \{0 < \Sigma_{i=1}^\rho |z_i|^{2m} < 1,\ \tau < \Sigma_{i=1}^k |w_i|^{2m} < 1\}.$$

Let

$$R = \{0 < \Sigma_{i=1}^{\rho}|z_i|^{2m} < 1,\ \Sigma_{i=1}^{k}|w_i|^{2m} < 1\}.$$

For $0 < \alpha < 1$, introduce

$$\varphi_\alpha(z,w) = (1-\alpha)\ \Sigma_{i=1}^{k}|w_i|^{2m} + \alpha(\Sigma_{i=1}^{\rho}|z_i|^{2m})^{-1} - 1.$$

By Lemma (10.1), φ_α is strongly ρ-convex on $\mathbb{C}^{\rho+k} \cap \{(z_1,\ldots,z_\rho) \neq 0\}$. Let $U_\alpha = R \cap \{\varphi_\alpha > 0\}$. Denote by ∂U_α the boundary of U_α in R. The following properties are easily verified:

(i) $U_\alpha \subset U_\beta$ for $\alpha < \beta$.

(ii) $R = \cup_{0<\alpha<1} U_\alpha$.

(iii) $\partial U_\alpha - R_\tau$ is compact.

(10.2.1) <u>Lemma</u>. Suppose V is a subvariety in $U_\alpha \cap R_\tau$ and V is disjoint from $R_{\tau'}$ for some $\tau' > \tau$. Then $\dim V \leqq \rho$.

Proof. We can assume that $V \neq \emptyset$. Take arbitrarily $(z^0,w^0) \in V$. Let $V_1 = V \cap \{z = z^0\}$. We need only show that $\dim V_1 \leqq 0$. We distinguish between two cases.

Case I. $\Sigma_{i=1}^{\rho}|z_i^0|^{2m} \leqq \alpha(1-\tau+\alpha\tau)^{-1}$. We have

$$(U_\alpha \cap R_\tau) \cap \{z = z^0\} = R_\tau \cap \{z = z^0\}.$$

Hence V_1 is a subvariety in $R_\tau \cap \{z = z^0\}$ and is disjoint from $R_{\tau'} \cap \{z = z^0\}$. Now apply Lemma (6.4) with $\varphi = \Sigma_{i=1}^k |w_i|^{2m}$, $c = \tau$,

$$G = \{\Sigma_{i=1}^k |w_i|^{2m} < \tau'\} \cap \{z = z^0\},$$

and

$$\tilde{G} = \{\Sigma_{i=1}^k |w_i|^{2m} < 1\} \cap \{z = z^0\}.$$

We conclude that $\dim V_1 \leqq 0$.

Case II. $\Sigma_{i=1}^\rho |z_i^0|^{2m} > \alpha(1-\tau+\alpha\tau)^{-1}$. We have

$$(U_\alpha \cap R_\tau) \cap \{z = z^0\} = U_\alpha \cap \{z = z^0\}.$$

Hence V_1 is a subvariety in $U_\alpha \cap \{z = z^0\}$ and is disjoint from $R_{\tau'} \cap \{z = z^0\}$. Now apply Lemma (6.4) with $\varphi = \varphi_\alpha$ and $c = 0$ and with G and $\tilde{G}$ the same as in Case I. We conclude that $\dim V_1 \leqq 0$. Q.E.D.

(10.2.2) <u>Lemma</u>. Suppose $\alpha < \alpha'$ and $\tau < \tau'$. Suppose $\mathcal{G}$ is a coherent analytic sheaf on R, $\mathcal{F}$ is a coherent analytic subsheaf of $\mathcal{G}$ on $U_\alpha \cup R_\tau$, and $\mathcal{F}'$ is a coherent analytic subsheaf of $\mathcal{G}$ on $U_{\alpha'} \cup R_{\tau'}$. Suppose $\mathcal{F}_\rho = \mathcal{F}$ and $\mathcal{F}'_\rho = \mathcal{F}'$. If $\mathcal{F}$ agrees with $\mathcal{F}'$ on $U_\alpha \cup R_{\tau'}$, then $\mathcal{F}$ agrees with $\mathcal{F}'$ on $(U_\alpha \cup R_\tau) \cap (U_{\alpha'} \cup R_{\tau'})$.

Proof. Suppose the contrary. Let V be the subvariety in $(U_\alpha \cup R_\tau) \cap (U_{\alpha'} \cup R_{\tau'})$ where $\mathcal{F}$ and $\mathcal{F}'$ disagree. Since $\mathcal{F}_\rho = \mathcal{F}$ and $\mathcal{F}'_\rho = \mathcal{F}'$, every branch of V has dimension $\geqq \rho + 1$. Since V is disjoint from $U_\alpha \cup R_{\tau'}$, V is contained in $U_{\alpha'} \cap R_\tau$. By Lemma (10.2.1), it follows from $V \cap R_{\tau'} = \emptyset$ that $\dim V \leqq \rho$. Contradiction.

Q.E.D.

(10.2.3) Lemma. Suppose $0 < \alpha < 1$, $0 < \tau < 1$, and $\mathcal{G}$ is a coherent analytic sheaf on R. Suppose $\mathcal{F}$ and $\mathcal{F}'$ are coherent analytic subsheaves of $\mathcal{G}$ on $U_\alpha \cup R_\tau$ such that $\mathcal{F}_\rho = \mathcal{F}$ and $\mathcal{F}'_\rho = \mathcal{F}'$. If $\mathcal{F}$ agrees with $\mathcal{F}'$ on R_τ, then $\mathcal{F}$ agrees with $\mathcal{F}'$ on $U_\alpha \cup R_\tau$.

Proof. Suppose the contrary. Let V be the subvariety in $U_\alpha \cup R_\tau$ where $\mathcal{F}$ and $\mathcal{F}'$ disagree. Since $\mathcal{F}_\rho = \mathcal{F}$ and $\mathcal{F}'_\rho = \mathcal{F}'$, every branch of V has dimension $\geqq \rho + 1$. Since V is disjoint from R_τ, $V \subset U_\alpha$. By Lemma (10.2.1), $\dim V \cap R_{\tau'} \leqq \rho$ for $0 < \tau' < \tau$. Hence $\dim V \leqq \rho$. Contradiction. Q.E.D.

(10.2.4) Lemma. Suppose $\mathcal{G}$ is a coherent analytic sheaf on R and $\mathcal{F}$ is a coherent analytic subsheaf of $\mathcal{G}$ on $U_\alpha \cup R_\tau$ such that $\mathcal{F}_\rho = \mathcal{F}$. Then for some $\alpha < \alpha' < 1$ there exists uniquely a coherent analytic subsheaf $\tilde{\mathcal{F}}$ of $\mathcal{G}$ on $U_{\alpha'} \cup R_\tau$ such that $\tilde{\mathcal{F}}$ agrees with $\mathcal{F}$ on $U_\alpha \cup R_\tau$ and satisfies $\tilde{\mathcal{F}}_\rho = \tilde{\mathcal{F}}$.

Proof. By Theorem (9.3), for every $x \in \partial U_\alpha - R_\tau$ we can extend $\mathcal{F}$ coherently to an open neighborhood of x as a subsheaf of $\mathcal{G}$. By the compactness of $\partial U_\alpha - R_\tau$, we can find open subsets W_i of R ($1 \leqq i \leqq m$) such that

(i) $\partial U_\alpha - R_\tau \subset \bigcup_{i=1}^m W_i$, and

(ii) there exists a coherent analytic subsheaf $\mathcal{F}^{(i)}$ of $\mathcal{G}$ on $U_\alpha \cup W_i$ extending $\mathcal{F}|U_\alpha$ and satisfying $\mathcal{F}_\rho^{(i)} = \mathcal{F}^{(i)}$.

Choose $\tau' > \tau$ and a relatively compact open subset W_i' of W_i ($1 \leqq i \leqq m$) such that $\partial U_\alpha - R_{\tau'} \subset \bigcup_{i=1}^m W_i'$. There exists $\alpha^* > \alpha$ such that $U_{\alpha^*} \subset R_\tau \cup (\bigcup_{i=1}^m W_i')$. Let $W_0 = R_\tau$, $W_0' = R_{\tau'}$, and $\mathcal{F}^{(0)} = \mathcal{F}$.

Let V_{ij} be the subvariety of $W_i \cap W_j$ where $\mathcal{F}^{(i)}$ and $\mathcal{F}^{(j)}$ disagree. Every branch of V_{ij} has dimension $\geqq \rho + 1$, because $\mathcal{F}_\rho^{(i)} = \mathcal{F}^{(i)}$ and $\mathcal{F}_\rho^{(j)} = \mathcal{F}^{(j)}$. Since V_{ij} is disjoint from U_α, according to Proposition (6.7) we have $V_{ij} \cap \partial U_\alpha = \emptyset$. Since $W_i' \cap W_j' \subset\subset W_i \cap W_j$ for $i \neq j$, there exists $\alpha^* > \alpha' > \alpha$ such that $V_{ij} \cap W_i' \cap W_j'$ is disjoint from $U_{\alpha'}$ for $0 \leqq i, j \leqq m$.

Define a coherent analytic subsheaf $\mathcal{F}'$ of $\mathcal{G}$ on $U_{\alpha'} \cup R_{\tau'}$ as follows:

$$\mathcal{F}' = \mathcal{F} \text{ on } U_\alpha$$

and

$$\mathcal{F}' = \mathcal{F}^{(i)} \text{ on } (U_{\alpha'} \cup R_{\tau'}) \cap W_i' \quad (0 \leqq i \leqq m).$$

$\mathcal{F}'$ is well defined, because

(i) $U_{\alpha'} \cup R_{\tau'} \subset U_\alpha \cup (\cup_{i=0}^m W_i')$,

(ii) $V_{ij} \cap W_i' \cap W_j'$ is disjoint from $U_{\alpha'}$, and

(iii) $\mathcal{F}^{(i)}$ extends $\mathcal{F} | U_\alpha$.

Clearly $\mathcal{F}'$ agrees with $\mathcal{F}$ on $U_\alpha \cup R_{\tau'}$. By Lemma (10.2.2), $\mathcal{F}'$ agrees with $\mathcal{F}$ on $(U_{\alpha'} \cup R_{\tau'}) \cap (U_\alpha \cup R_\tau)$.

Define a coherent analytic subsheaf $\tilde{\mathcal{F}}$ of $\mathcal{G}$ on $U_\alpha \cup R_\tau$ as follows:

$$\tilde{\mathcal{F}} = \mathcal{F} \text{ on } U_\alpha \cup R_\tau$$

and

$$\tilde{\mathcal{F}} = \mathcal{F}' \text{ on } U_{\alpha'} \cup R_{\tau'} .$$

$\tilde{\mathcal{F}}$ extends $\mathcal{F}$ and satisfies $\tilde{\mathcal{F}}_\rho = \tilde{\mathcal{F}}$. By Lemma (10.2.3), $\tilde{\mathcal{F}}$ is unique. Q.E.D.

(10.2.5) <u>Proposition</u>. Suppose $\mathcal{G}$ is a coherent analytic sheaf on R and $\mathcal{F}$ is a coherent analytic subsheaf of $\mathcal{G}$ on $U_{\alpha_0} \cup R_\tau$ such that $\mathcal{F}_\rho = \mathcal{F}$. Then $\mathcal{F}$ can be extended uniquely to a coherent analytic subsheaf $\tilde{\mathcal{F}}$ of $\mathcal{G}$ on R satisfying $\tilde{\mathcal{F}}_\rho = \tilde{\mathcal{F}}$.

Proof. Let A be the set of points α in $[\alpha_0, 1)$ such that $\mathcal{F}$ can be extended to a coherent analytic subsheaf $\mathcal{F}^{(\alpha)}$ of $\mathcal{G}$ on $U_\alpha \cup R_\tau$ satisfying $(\mathcal{F}^{(\alpha)})_\rho = \mathcal{F}^{(\alpha)}$. By Lemma (10.2.3), $\mathcal{F}^{(\alpha)}$ is unique.

Let β be the supremum of A. We claim that $\beta = 1$. Suppose the contrary. There exists an increasing sequence $\{\beta_i\} \subset A$ converging to β. Define a coherent analytic subsheaf $\mathcal{F}^{(\beta)}$ of $\mathcal{G}$ on $U_\beta \cup R_\tau$ as follows:

$$\mathcal{F}^{(\beta)} = \mathcal{F}^{(\beta_i)} \text{ on } U_{\beta_i} \cup R_\tau .$$

Then $\mathcal{F}^{(\beta)}$ is well defined. It extends $\mathcal{F}$ and satisfies $(\mathcal{F}^{(\beta)})_\rho = \mathcal{F}^{(\beta)}$. By Lemma (10.2.4) there exists $\beta < \beta' < 1$ such that $\mathcal{F}^{(\beta)}$ can be extended to a coherent analytic subsheaf $\mathcal{F}^{(\beta')}$ of $\mathcal{G}$ on $U_{\beta'} \cup R_\tau$ satisfying $(\mathcal{F}^{(\beta')})_\rho = \mathcal{F}^{(\beta')}$. Hence $\beta' \in A$, contradicting $\beta = \sup A$. The claim is proved.

Define a coherent analytic subsheaf $\tilde{\mathcal{F}}$ of $\mathcal{G}$ on R as follows:

$$\tilde{\mathcal{F}} = \mathcal{F}^{(\alpha)} \text{ on } U_\alpha \cup R_\tau \text{ for } \alpha \in A.$$

$\tilde{\mathcal{F}}$ satisfies the requirement. Q.E.D.

(10.3.1) <u>Lemma</u>. Suppose D is an open subset of $\mathbb{C}^\rho$, $0 < \beta_i' < \beta_i$ $(1 \leq i \leq k)$, and $\mathcal{G}$ is a coherent analytic sheaf on $D \times K^k(\beta_1, \ldots, \beta_k)$. Suppose $\mathcal{F}$ and $\mathcal{F}'$ are coherent analytic subsheaves of $\mathcal{G}$ such that $\mathcal{F}_\rho = \mathcal{F}$ and $\mathcal{F}'_\rho = \mathcal{F}'$. If $\mathcal{F}$ and $\mathcal{F}'$ agree on

$D \times G^k(\beta_1',\ldots,\beta_k';\beta_1,\ldots,\beta_k)$, then $\mathcal{F}$ and $\mathcal{F}'$ agree on $D \times K^k(\beta_1,\ldots,\beta_k)$.

Proof. Suppose the contrary. Let V be the subvariety in $D \times K^k(\beta_1,\ldots,\beta_k)$ where $\mathcal{F}$ and $\mathcal{F}'$ disagrees. Since $\mathcal{F}_\rho = \mathcal{F}$ and $\mathcal{F}_\rho' = \mathcal{F}'$, every branch of V has dimension $\geqq \rho + 1$. Take $x \in V$. Let $V_1 = V \cap \{z_i = z_i(x) \text{ for } 1 \leqq i \leqq \rho\}$. Then $\dim V_1 \geqq 1$. However, since V is disjoint from $D \times G^k(\beta_1',\ldots,\beta_k';\beta_1,\ldots,\beta_k)$, V_1 is compact and hence $\dim V_1 \leqq 0$. Contradiction. Q.E.D.

(10.3.2) Theorem. Suppose $0 < \alpha_i' < \alpha_i$ $(1 \leqq i \leqq \rho)$ and $0 < \beta_j' < \beta_j$ $(1 \leqq j \leqq k)$. Let

$$K = K^{\rho+k}(\alpha_1,\ldots,\alpha_\rho,\beta_1,\ldots,\beta_k)$$

and let H be the union of

$$K^\rho(\alpha_1,\ldots,\alpha_\rho) \times G^k(\beta_1',\ldots,\beta_k';\beta_1,\ldots,\beta_k)$$

and

$$K^{\rho+k}(\alpha_1',\ldots,\alpha_\rho',\beta_1,\ldots,\beta_k).$$

Suppose $\mathcal{G}$ is a coherent analytic sheaf on K and $\mathcal{F}$ is a coherent analytic subsheaf of $\mathcal{G}$ on H such that $\mathcal{F}_\rho = \mathcal{F}$. Then $\mathcal{F}$ can be extended uniquely to a coherent analytic subsheaf $\mathcal{F}'$ of $\mathcal{G}$ on K satisfying $\mathcal{F}'_\rho = \mathcal{F}'$.

Proof. We can assume w.l.o.g. that $\alpha_i = 1$ and $\beta'_j < 1 < \beta_j$. We can find a natural number m_0 such that for $m \geqq m_0$ there exists $\tau = \tau(m)$ satisfying

$$R_\tau \subset K^\rho(1,\ldots,1) \times G^k(\beta'_1,\ldots,\beta'_k;1,\ldots,1),$$

where R_τ has the same meaning as in (10.2).

Fix arbitrarily $m \geqq m_0$. There exists $0 < \sigma < 1$ such that $U_\sigma \subset H$. By Proposition (10.2.5) we can extend $\mathcal{F}|U_\sigma \cup R_\tau$ to a coherent analytic subsheaf $\mathcal{F}^{(m)}$ of $\mathcal{G}$ on R which satisfies $\mathcal{F}^{(m)}_\rho = \mathcal{F}^{(m)}$, where R has the same meaning as in (10.2). Define a coherent analytic subsheaf $\tilde{\mathcal{F}}^{(m)}$ of $\mathcal{G}$ on

$$G_m = \{\Sigma_{i=1}^\rho |z_i|^{2m} < 1\} \cap \{|w_j| < \beta_j \text{ for } 1 \leqq j \leqq k\}$$

as follows:

$$\tilde{\mathcal{F}}^{(m)} = \mathcal{F}^{(m)} \text{ on } R$$

and

$$\begin{cases} \tilde{\mathcal{F}}^{(m)} = \mathcal{F} \text{ on } G_m \cap \{\tau < \Sigma_{j=1}^k |w_j|^{2m}\} \\ \text{and } \{\Sigma_{i=1}^\rho |z_i|^{2m} < \sigma\} \cap \{|w_j| < \beta_j \text{ for } 1 \leqq j \leqq k\}. \end{cases}$$

$\{G_m\}$ is an increasing sequence of open subsets of K whose union is K. Since $\tilde{\mathcal{F}}^{(m)}$ and $\tilde{\mathcal{F}}^{(m+1)}$ agree on

$$\{\Sigma_{i=1}^{\rho} |z_i|^{2m} < 1\} \cap (\cup_{j=1}^{k}\{1<|w_j|< \beta_j, |w_\ell| < \beta_\ell \text{ for } 1\leqq \ell \leqq k\}),$$

by Lemma (10.3.1), $\tilde{\mathcal{F}}^{(m)}$ and $\tilde{\mathcal{F}}^{(m+1)}$ agree on G_m. Define a coherent analytic subsheaf $\tilde{\mathcal{F}}$ of $\mathcal{G}$ on K as follows:

$$\tilde{\mathcal{F}} = \tilde{\mathcal{F}}^{(m)} \text{ on } G_m \text{ for } m \geqq m_0.$$

Then $\tilde{\mathcal{F}}$ satisfies the requirement. Q.E.D.

(10.4) Suppose X is a complex space and $\varphi: X \to (a,b)$ is a C^∞ proper map (where $a \in \{-\infty\} \cup \mathbb{R}$ and $b \in \mathbb{R} \cup \{\infty\}$) such that φ is strongly ρ-convex. For $a \leqq c < b$ let $X_c = \{\varphi > c\}$.

(10.4.1) <u>Lemma</u>. Suppose $a \leqq c < b$ and $\mathcal{G}$ is a coherent analytic sheaf on X. Suppose $\mathcal{F}$ and $\mathcal{F}'$ are coherent analytic subsheaves of $\mathcal{G}$ on X_c such that $\mathcal{F}_{\rho-1} = \mathcal{F}$ and $\mathcal{F}'_{\rho-1} = \mathcal{F}'$. If $\mathcal{F}$ and $\mathcal{F}'$ agree on $X_{c'}$ for some $c' > c$, then $\mathcal{F}$ and $\mathcal{F}'$ agree on X_c.

Proof. Suppose the contrary. Let V be the subvariety of X_c where $\mathcal{F}$ and $\mathcal{F}'$ disagree. Since $\mathcal{F}_{\rho-1} = \mathcal{F}$ and $\mathcal{F}'_{\rho-1} = \mathcal{F}'$, every branch of V has dimension $\geqq \rho$. Take $x \in V$. The supremum of φ on V is equal to the supremum of φ on the compact set $V \cap \{\varphi(x_0) \leqq \varphi \leqq c'\}$. φ achieves its supremum on V, contradicting Proposition (6.6). Q.E.D.

(10.4.2) Lemma. Suppose $\mathcal{G}$ is a coherent analytic sheaf on X. Suppose $a < c < b$ and $\mathcal{F}$ is a coherent analytic subsheaf on X_c such that $\mathcal{F}_\rho = \mathcal{F}$. Then for some $a < c' < c$ there exists uniquely a coherent analytic subsheaf $\tilde{\mathcal{F}}$ of $\mathcal{G}$ on $X_{c'}$ such that $\tilde{\mathcal{F}}$ agrees with $\mathcal{F}$ on X_c and satisfies $\tilde{\mathcal{F}}_\rho = \tilde{\mathcal{F}}$.

Proof. By Theorem (10.3), for every $x \in \{\varphi = c\}$ we can extend $\mathcal{F}$ coherently to an open neighborhood of x as a subsheaf of $\mathcal{G}$. By the compactness of $\{\varphi=c\}$, we can find open subsets W_i of $X (1 \leq i \leq m)$ such that

(i) $\{\varphi = c\} \subset \bigcup_{i=1}^m W_i$, and

(ii) there exists a coherent analytic subsheaf $\mathcal{F}^{(i)}$ of $\mathcal{G}$ on $X_c \cup W_i$ and satisfying $\mathcal{F}^{(i)}_\rho = \mathcal{F}^{(i)}$.

Choose a relatively compact open subset W_i' of W_i $(1 \leq i \leq m)$ such that $\{\varphi = c\} \subset \bigcup_{i=1}^m W_i'$. There exists $a < c^* < c$ such that $X_{c^*} \subset X_c \cup (\bigcup_{i=1}^m W_i')$. Let V_{ij} be the subvariety of $W_i \cap W_j$ where $\mathcal{F}^{(i)}$ and $\mathcal{F}^{(j)}$ disagree. Every branch of V_{ij} has dimension $\geq \rho + 1$, because $\mathcal{F}^{(i)}_\rho = \mathcal{F}^{(i)}$ and $\mathcal{F}^{(j)}_\rho = \mathcal{F}^{(j)}$. Since $V_{ij} \cap X_c = \emptyset$, by Proposition (6.7) we have $V_{ij} \cap \{\varphi = c\} = \emptyset$. Since $W_i' \cap W_j' \subset\subset W_i \cap W_j$, there exists $c^* < c' < c$ such that $V_{ij} \cap W_i' \cap W_j'$ is disjoint from $X_{c'}$ for $1 \leq i, j \leq m$.

Define a coherent analytic subsheaf $\tilde{\mathcal{F}}$ on $X_{c'}$ as follows:

$$\tilde{\mathcal{F}} = \mathcal{F} \text{ on } X_c$$

and

$$\tilde{\mathcal{F}} = \mathcal{F}^{(i)} \text{ on } X_{c'} \cap W_i' .$$

$\tilde{\mathcal{F}}$ is well defined. It extends $\mathcal{F}$ and satisfies $\tilde{\mathcal{F}}_\rho = \tilde{\mathcal{F}}$. The uniqueness of $\tilde{\mathcal{F}}$ follows from Lemma (10.4.1). Q.E.D.

(10.4.3) Theorem. Suppose $\mathcal{G}$ is a coherent analytic sheaf on X. Suppose $a < c_0 < b$ and $\mathcal{F}$ is a coherent analytic subsheaf of $\mathcal{G}$ on X_{c_0} such that $\mathcal{F}_\rho = \mathcal{F}$. Then $\mathcal{F}$ can be extended uniquely to a coherent analytic subsheaf $\tilde{\mathcal{F}}$ of $\mathcal{G}$ on X satisfying $\tilde{\mathcal{F}}_\rho = \tilde{\mathcal{F}}$.

Proof. Let C be the set of points c in $(a, c_0]$ such that $\mathcal{F}$ can be extended to a coherent analytic subsheaf $\mathcal{F}^{(c)}$ of $\mathcal{G}$ on X_c satisfying $(\mathcal{F}^{(c)})_\rho = \mathcal{F}^{(c)}$. By Lemma (10.4.1), $\mathcal{F}^{(c)}$ is unique.

Let $\tilde{c}$ be the infemum of C. We claim that $\tilde{c} = a$. Suppose the contrary. There exists a decreasing sequence $\{c_i\} \subset C$ converging to $\tilde{c}$. Define a coherent analytic subsheaf $\mathcal{F}^{(\tilde{c})}$ of $\mathcal{G}$ on $X_{\tilde{c}}$ as follows:

$$\mathcal{F}^{(\tilde{c})} = \mathcal{F}^{(c_i)} \text{ on } X_{c_i} .$$

Then $\mathcal{F}^{(\tilde{c})}$ is well defined. It extends $\mathcal{F}$ and satisfies $(\mathcal{F}^{(\tilde{c})})_\rho = \mathcal{F}^{(\tilde{c})}$. By Lemma (10.4.2) there exists $a<c'<\tilde{c}$ such that $\mathcal{F}^{(\tilde{c})}$ can be extended to a coherent analytic subsheaf $\mathfrak{F}^{(c')}$ of $\mathcal{G}$ on $X_{c'}$ satisfying $(\mathfrak{F}^{(c')})_\rho = \mathfrak{F}^{(c')}$. Hence $c' \in C$, contradicting $c = \inf C$. The claim is proved.

Define a coherent analytic subsheaf $\tilde{\mathfrak{F}}$ of $\mathcal{G}$ on X as follows:

$$\tilde{\mathfrak{F}} = \mathfrak{F}^{(c)} \text{ for } c \in C.$$

$\tilde{\mathfrak{F}}$ satisfies the requirement. Q.E.D.

Historical notes

§1

The theory of singularities of coherent sheaves and their cohomology classes was started by G. Scheja. Theorem (1.11) is due to him [20] and he proved also (b) => (d) of Theorem (1.14) [19]. The rest of Theorem (1.14) is due to Trautmann [36] [37].

§2

Primary decomposition and relative gap-sheaves were first considered by Thimm. Theorems (2.2) and (2.7), except for the relation $\mathcal{F}_d = \mathcal{F}[S_d(\mathcal{G}/\mathcal{F})]$ in Theorem (2.7), are due to him [32]. However, the simple proofs for these theorems given here are completely different from his and, to the knowledge of the authors, are new. Theorems (2.12) and (2.14) were proved by Spallek [30] and the second part of Theorem (2.14) also by Siu [26], but the proofs given here are new.

§3

The method of applying the Hom functor to a projective resolution of $\mathcal{F}^*$ to get an injective resolution of $\mathcal{F}$ was first introduced by Trautmann [36], where he obtained Corollary (3.6). The method of annihilating undesirable local cohomology groups by powers of ideal sheaf sections was introduced by Siu [27]. Theorem (3.5) is due to Trautmann [37] and Siu [27]. The proof of Theorem (3.5) given here is adopted from Siu-Trautmann [29]. Absolute gap-sheaves of the form $\mathcal{R}_d^0\mathcal{F}$ were introduced by Siu [25], where he proved (a) <=> (d) of Corollary (3.9).

§4

All the results in §4 are due to Siu-Trautmann [29] and the proofs given here are the same as those given in [29].

§5

Theorem (5.8) is due to Serre [23]. Theorem (5.9) is due to Martineau [16]. Theorem (5.12) is due to Harvey [11]. Corollary (5.13) is due to Suominen [31].

§6–§8

The concept of ρ-convexity was first introduced by Rothstein to study the local extension of subvarieties and Theorem (8.3) is due to him [18]. Though the techniques of representing subvarieties locally as analytic covers to solve extension problems are well-known, Theorem (8.4) for $k = 1$ was first formulated by Fujimoto [5].

§9

Thimm is the first one to use relative gap-sheaves and meromorphic functions to investigate coherent subsheaf extension [33]. Theorem (9.3) is due to Siu-Trautmann [28] and it was proved there by methods of normalization and absolute gap-sheaves. The proof of Theorem (9.3) presented here is new.

§10

The exhaustion techniques used here are well-known. The globalization proof presented here is a refinement of §8 of [18].

Bibliography

1. Andreotti, A., and Grauert, H., "Théorèmes de finitude pour la cohomologie des espaces complexes". Bull. Soc. Math. France 90, 193-259 (1962).

2. Cartan, H., "Faisceaux analytiques cohérents". C.I.M.E., 1963, Inst. Math. d. Univ., Roma, 1-88.

3. Dolbeault, P., "Sur la cohomologie des variétés analytiques complexes". C.R. Acad. Sci. Paris 236, 175-177 (1953).

4. Frenkel, J., "Cohomologie nonabélienne et espaces fibrés". Bull. Soc. Math. France 83, 135-218 (1957).

5. Fujimoto, H., "On the continuation of analytic sets". J. Math. Soc. Japan 18, 51-85 (1966).

6. Godement, R., Topologie algébrique et théorie des faisceaux. Paris: Hermann, 1964.

7. Grauert, H., Ein Theorem der analytischen Garbentheorie und die Modulräume komplexer Strukturen. I.H.E.S. (Paris), No. 5 (1960).

8. Grothendieck, A., "Cohomologie locale des faisceaux cohérents et théorèmes de Lefschetz locaux et globaux". Séminaire de géométrie algébrique. I.H.E.S. (Paris), 1962.

9. Gunning, R.C., and Rossi, H., Analytic functions of several complex variables. Englewood Cliffs, N.J.: Prentice-Hall, 1965.

10. Hartshorne, R., Local cohomology. Lecture Notes No. 41, Berlin-Heidelberg-New York: Springer, 1967.

11. Harvey, R., "The theory of hyperfunctions on totally real subsets of a complex manifold with applications to extension problems". Amer. J. Math. 91, 853-873 (1969).

12. Hörmander, L., An introduction to complex analysis in several variables. Princeton, N.J.: Van Nostrand, 1966.

13. Horvath, J., Topological vector spaces and distributions I. Reading, Mass.: Addison-Wesley, 1966.

14. Malgrange, B., "Faisceaux sur des variétés analytiques-réelles." Bull. Soc. Math. France 85, 231-237 (1957).

15. Malgrange, B., Analytic spaces. Monographie No. 17. L'Enseignement Math. Geneve, 1968.

16. Martineau, A., "Les hyperfonctions de M. Sato". Séminaire Bourbaki 1960-61, Expose 214.

17. Narasimhan, R., Introduction to the theory of analytic spaces. Lecture Notes No. 25, Berlin-Heidelberg-New York: Springer, 1966.

18. Rothstein, W., "Zur Theorie der analytischen Mannifaltigkeiten in Raume von n komplexen Veränderlichen". Math Ann. 129, 96-138 (1955).

19. Scheja, G., "Riemannsche Hebbarkeitssätze für Cohomologieklassen". Math. Ann. 144, 345-360 (1962).

20. Scheja, G., "Fortsetzungssätze der komplex-analytischen Cohomologie und ihre algebraische Charakterisierung". Math. Ann. 157, 75-94 (1964).

21. Serre, J.-P., "Un théorème de dualité." Comm. Math. Helv. 29, 9-26 (1955).

22. Serre, J.-P., Algèbre locale - multiplicités. Lecture Notes No. 11, Berlin-Heidelberg-New York: Springer, 1965.

23. Serre, J.-P., "Prolongement de faisceaux analytiques cohérents". Ann. Inst. Fourier 16, 363-374 (1966).

24. Siu, Y.-T., "Noether-Lasker decomposition of coherent analytic subsheaves". Trans. Amer. Math. Soc. 135, 375-384 (1969).

25. Siu, Y.-T., "Absolute gap-sheaves and extensions of coherent analytic sheaves". Trans. Amer. Math. Soc. 141, 361-376 (1969).

26. Siu, Y.-T., "O^N-approximable and holomorphic functions on complex spaces". Duke Math. J. 36, 451-454 (1969).

27. Siu, Y.-T., "Analytic sheaves of local cohomology". Trans. Amer. Math. Soc. 148, 347-366 (1970).

28. Siu, Y.-T., and Trautmann, G., "Extension of coherent analytic subsheaves". Math. Ann. 188, 128-142 (1970).

29. Siu, Y.-T., and Trautmann, G., "Closedness of coboundary modules of analytic sheaves". Trans. Amer. Math. Soc. (to appear).

30. Spallek, K.-H., "Zum Satz von Osgood und Hartogs für analytische Moduln I". Math. Ann. 178, 83-118 (1968).

31. Suominen, K., "Duality for coherent sheaves on analytic manifolds". Ann. Acad. Sci. Fennical A.I. 424, 1-19 (1968).

32. Thimm. W., "Lückengarben von kohärenten analytischen Modulgarben". Math. Ann. 148, 372-394 (1962).

33. Thimm, W., "Struktur- und Singularitätsuntersuchungen an kohärenten analytischen Modulgarben". J. reine u. angew. Math. 234, 123-151 (1969).

34. Thimm, W., "Fortsetzung von kohärenten analytischen Modulgarben". Math. Ann. 184, 329-353 (1970).

35. Trautmann, G., "Eine Bemerkung zur Struktur der kohärenten analytischen Garben". Arch. Math. 19, 300-304 (1968).

36. Trautmann, G., "Cohérence de faisceaux analytiques de la cohomologie locale". C. R. Acad. Sci. Paris 267, 694-695 (1968).

37. Trautmann, G., "Ein Endlichkeitssatz in der analytischen Geometrie". Invent. math. 8, 143-174 (1969).